CATWATCHING
and
CATLORE

CATWATCHING
and
CATLORE

New enlarged edition

Desmond Morris

ARROW

Arrow Books Limited
20 Vauxhall Bridge Road, London SW1V 2SA

An imprint of Random House UK Limited

London Melbourne Sydney Auckland
Johannesburg and agencies throughout
the world

First published in Great Britain by Jonathan Cape

This edition published in 1992 by Arrow Books

1 3 5 7 9 10 8 6 4 2

Printed and bound in Great Britain by
Cox & Wyman Ltd, Reading, Berkshire

ISBN 0 09 922901 3

Contents

Introduction

The domestic cat is a contradiction. No animal has developed such an intimate relationship with mankind, while at the same time demanding and getting such independence of movement and action. The dog may be man's best friend, but it is rarely allowed out on its own to wander from garden to garden or street to street. The obedient dog has to be taken for a walk. The headstrong cat walks alone.

The cat leads a double life. In the home it is an overgrown kitten gazing up at its human owners. Out on the tiles it is fully adult, its own boss, a free-living wild creature, alert and self-sufficient, its human protectors for the moment completely out of mind. This switch from tame pet to wild animal and then back again is fascinating to watch. Any cat-owner who has accidentally come across their pet cat out-of-doors, when it is deeply involved in some feline soap opera of sex and violence, will know what I mean. One instant the animal is totally wrapped up in an intense drama of courtship or status. Then out of the corner of its eye it spots its human owner watching the proceedings. There is a schizoid moment of double involvement, a hesitation, and the animal runs across, rubs against its owner's leg and becomes the house-kitten once more.

The cat manages to remain a tame animal because of the sequence of its upbringing. By living both with other cats (its mother and litter-mates) and with humans (the family that has adopted it) during its infancy and kittenhood, it becomes attached to both and considers that it belongs to both species. It is like a child that grows up in a foreign country and as a consequence becomes bilingual. The cat becomes bi-mental. It may be a cat physically, but mentally it is both feline and human. Once it is fully adult, however, most of its responses are feline ones and it has only one major reaction to its human owners. It treats them as pseudo-parents. This is because they took over from the real mother at a sensitive stage of the kitten's development and went on giving it milk, solid food and comfort as it grew up.

This is rather different from the kind of bond that develops between man and dog. The dog does see its human owners as pseudo-parents, like the cat. On that score the process of attachment is similar. But the dog has an additional link. Canine society is group-organized, feline society is not. Dogs live in packs with tightly controlled status relationships between the individuals. There are top dogs, middle dogs and bottom dogs, and under natural circumstances they move around together, keeping tabs on one another the whole time. So the adult pet dog sees its human family both as pseudo-parents and as the dominant members of its pack. Hence its renowned reputation for obedience and its celebrated capacity for loyalty. Cats do have a complex social organization, but they never hunt in packs. In the wild most of their day is spent in solitary stalking. Going for a walk with a human therefore has no appeal for them. And as for 'coming to heel' and learning to 'sit' and 'stay', they are simply not interested. Such manoeuvres have no meaning for them.

So the moment a cat manages to persuade a human being to open a door (that most hated of human inventions), it is off and away without a backward glance. As it crosses the threshold the cat becomes transformed. The kitten-of-man brain is switched off and the wild-cat brain is clicked on. The dog, in such a situation, may look back to see if its human pack-mate is following to join in the fun of exploring, but not the cat. The cat's mind has floated off into another, totally feline world, where strange bipedal apes have no place.

Because of this difference between domestic cats and domestic dogs, cat-lovers tend to be rather different from dog-lovers. As a rule they have a stronger personality-bias towards independent thought and action. Artists like cats; soldiers like dogs. The much-lauded 'group loyalty' phenomenon is alien to both cats and cat-lovers. If you are a company man, a member of the gang, one of the lads, or picked for the squad, the chances are that at home there is no cat curled up in front of the fire. The ambitious Yuppie, the aspiring politician, the professional footballer, these are not typical cat-owners. It is difficult to picture a rugger-player with a cat in his lap –

much easier to envisage him taking his dog for a walk.

Those who have studied cat-owners and dog-owners as two distinct groups report that there is also a gender-bias. Cat-lovers show a greater tendency to be female. This is not surprising in view of the division of labour that developed during human evolution. Prehistoric males became specialized as group hunters, while the females concentrated on food-gathering and child-rearing. This difference led to a human male 'pack mentality' that is far less marked in females. Wolves, the wild ancestors of domestic dogs, also became pack-hunters, so the modern dog has much more in common with the human male than the human female. An anti-female commentator could refer to women and cats as lacking in team-spirit; an anti-male one to men and dogs as gangsters.

The argument will always go on — feline self-sufficiency and individualism versus canine camaraderie and good-fellowship. But it is important to stress that in making a valid point I have caricatured the two positions. In reality there are many people who enjoy equally the company of both cats and dogs. And all of us, or nearly all of us, have both feline and canine elements in our personalities. We have moods when we want to be alone and thoughtful, and other times when we wish to be in the centre of a crowded, noisy room.

Both the cat and the dog are animals with which we humans have entered into a solemn contract. We made an unwritten, unspoken pact with their wild ancestors, offering food and drink and protection in exchange for the performance of certain duties. For dogs, the duties were complex, involving a whole range of hunting tasks, as well as guarding property, defending their owners against attack, destroying vermin, and acting as beasts of burden pulling our carts and sledges. In modern times an even greater range of duties has been given to the patient, long-suffering canine, including such diverse activities as guiding the blind, trapping criminals and running races.

For cats, the terms of the ancient contract were much simpler and have always remained so. There was just one primary task and one secondary one. They were required to act firstly as pest-controllers and then, in addition, as household pets. Because they are solitary hunters of small prey they were of little use to human huntsmen in the field. Because they do not live in tightly organized social groups depending on mutual aid to survive, they do not raise the alarm in response to intruders in the home, so they were little use as guardians of property, or as defenders of their owners. Because of their small size they could offer no assistance whatever as beasts of burden. In modern times, apart from sharing the honours with dogs as the ideal house-pets, and occasionally sharing the acting honours in films and plays, cats have failed to diversify their usefulness to mankind.

Despite this narrower involvement in human affairs, the cat has managed to retain its grip on human affections. There are almost as many cats as dogs in the British Isles, according to recent estimates: about five million cats to six million dogs. In the United States the ratio is slightly less favourable to felines: about twenty-three million cats to forty million dogs. Even so, this is a huge population of domestic cats and, if anything, it is probably an underestimate. Although there are still a few mousers and ratters about, performing their ancient duties as vermin destroyers, the vast majority of all domestic cats today are family pets or feral survivors. Of the family pets, some are pampered pedigrees but most are moggies of mixed ancestry. The proportion of pedigree cats to moggies is probably lower than that for pedigree dogs to mongrels. Although cat shows are just as fiercely contested as dog shows, there are fewer of them, just as there are fewer breeds of show cats. Without the wide spectrum of ancient functions to fulfil, there was far less breed specialization in the early days. Indeed, there was hardly any. All breeds of cat are good mousers and ratters, and no more was required of them. So any modifications in coat length, colour or pattern, or in body proportions, had to arise purely on the basis of local preferences and owners' whims. This has led to some strikingly beautiful pedigree cat breeds, but not to the amazing range of dramatically different types found among dogs. There is no cat equivalent of the Great Dane or the Chihuahua, the St Bernard or the Dachshund. There is a good deal of variation in fur type and colour, but very little in body shape and size. A really large cat weighs in at around eighteen pounds; the smallest at about three pounds. This means that, even when considering almost freakish feline

4

extremes, big domestic cats are only six times as heavy as small ones, compared with the situation among dogs, where a St Bernard can weigh 300 times as much as a little Yorkshire Terrier. In other words, the weight variation of dogs is fifty times as great as in cats.

Turning to abandoned cats and those that have gone wild through choice – the feral population – one also notices a considerable difference. Stray dogs form self-supporting packs and start to breed and fend for themselves without human aid in less civilized regions, but such groups have become almost non-existent in urban and suburban areas. Indeed, in modern, crowded European countries they hardly exist anywhere. Even the rural districts cannot support them. If a feral pack forms, it is soon hunted down by the farming community to prevent attacks on their stock. Feral cat colonies are another matter. Every major city has a thriving population of them. Attempts to eradicate them usually fail because there are always new strays to add to the pool. And the need to destroy them is not felt to be so great, because they can often survive by continuing their age-old function of pest control. Where human intervention has eliminated the rat and mouse population by poisoning, however, the feral cats must live on their wits, scavenging from dustbins and begging from soft-hearted humans. Many of these back-alley cats are pathetic creatures on the borderline of survival. Their resilience is amazing and a testimony to the fact that, despite millennia of domestication, the feline brain and body are still remarkably close to the wild condition.

At the same time this resilience is to blame for a great deal of feline suffering. Because cats *can* survive when thrown out and abandoned, it makes it easier for people to do just that. The fact that most of these animals must then live out their years in appalling urban conditions – slum cats scratching a living among the garbage and filth of human society – may reflect how tough they are, but it is a travesty of feline existence. That we tolerate it is one more example of the shameful manner in which we have repeatedly broken our ancient contract with the cat. It is nothing, however, compared with the brutal way we have sometimes tormented and tortured cats over the centuries. They have all too frequently been the butt of our redirected aggression, so much so that we even have a popular saying to express the phenomenon: '... and the office boy kicked the cat', illustrating the way in which insults from above become diverted to victims lower down the social order, with the cat at the bottom of the ladder.

Fortunately, against this can be set the fact that the vast majority of human families owning pet cats do treat their animals with care and respect. The cats have a way of endearing themselves to their owners, not just by their 'kittenoid' behaviour, which stimulates strong parental feelings, but also by their sheer gracefulness. There is an elegance and a composure about them that captivates the human eye. To the sensitive human being it becomes a privilege to share a room with a cat, exchange its glance, feel its greeting rub, or watch it gently luxuriate itself into a snoozing ball on a soft cushion. And for millions of lonely people – many physically incapable of taking long walks with a demanding dog – the cat is the perfect companion. In particular, for people forced to live on their own in later life, their company provides immeasurable rewards. Those tight-lipped puritans who, through callous indifference and a sterile selfishness, seek to stamp out all forms of pet-keeping in modern society would do well to pause and consider the damage their actions may cause.

This brings me to the purpose of *Catwatching*. As a zoologist, I have had in my care, at one time or another, most members of the cat family, from great Tigers to tiny Tiger-Cats, from powerful Leopards to diminutive Leopard-Cats, and from mighty Jaguars to rare little Jaguarondis. At home there has nearly always been a domestic moggie to greet my return, sometimes with a cupboard-full of kittens. As a boy growing up in the Wiltshire countryside, I spent many hours lying in the grass, observing the farm cats as they expertly stalked their prey, or spying on the hayloft nests where they suckled their squirming kittens. I developed the habit of catwatching early in life, and it has stayed with me now for nearly half a century. Because of my professional involvement with animals I have frequently been asked questions about cat behaviour, and I have been surprised at how little most people seem to know about these intriguing animals. Even those who dote on their own pet cat often have only a vague understanding of the complexities of its social life, its sexual behaviour, its aggression or its hunting skills. They know its moods well and care for it fastidiously, but they do not go out of their way to study their pet. To some extent this is not their fault, as much feline behaviour takes place away from the home base of the kitchen and the living room. So I hope that even those who feel they know their own cats intimately may learn a little more about their graceful companions by reading these pages.

The method I have used is to set out a series of basic questions and then to provide simple, straightforward answers to them. There are plenty of good, routine books on cat care, which give all the usual details about feeding, housing and veterinary treatment, combined with a classificatory list of the various cat breeds and their characteristics. I have not repeated those details here. Instead I have tried to provide a different sort of cat book, one that concentrates on feline behaviour and gives replies to the sort of queries with which I have been confronted over the years. If I have succeeded, then, the next time you encounter a cat, you should be able to view the world in a more feline way. And once you have started to do that, you will find yourself asking more and more questions about their fascinating world, and perhaps you too will develop the urge to do some serious catwatching.

The Cat

We know for certain that 3,500 years ago the cat was already fully domesticated. We have records from ancient Egypt to prove this. But we do not know when the process began. The remains of cats have been found at a neolithic site at Jericho dating from 9,000 years ago, but there is no proof that those felines were domesticated ones. The difficulty arises from the fact that the cat's skeleton changed very little during its shift from wild to tame. Only when we have specific records and detailed pictures – as we do from ancient Egypt – can we be sure that the transformation from wild cat to domestic animal had taken place.

One thing is clear: there would have been no taming of the cat before the Agricultural Revolution (in the neolithic period, or New Stone Age). In this respect the cat differed from the dog. Dogs had a significant role to play even before the advent of farming. Back in the palaeolithic period (or Old Stone Age), prehistoric human hunters were able to make good use of a four-legged hunting companion with superior scenting abilities and hearing. But cats were of little value to early man until he had progressed to the agricultural phase and was starting to store large quantities of food. The grain stores, in particular, must have attracted a teeming population of rats and mice almost from the moment that the human hunter settled down to become a farmer. In the early cities, where the stores were great, it would have become an impossible task for human guards to ambush the mice and kill them in sufficient numbers to stamp them out or even to prevent them from multiplying. A massive infestation of rodents must have been one of the earliest plagues known to urban man. Any carnivore that preyed on these rats and mice would have been a godsend to the harassed food-storers.

It is easy to visualize how one day somebody made the casual observation that a few wild cats had been noticed hanging around the grain stores, picking off the mice. Why not encourage them? For the cats, the scene must have been hard to believe. There all around them was a scurrying feast on a scale they had never encountered before. Gone were the interminable waits in the undergrowth. All that was needed now was a leisurely stroll in the vicinity of the vast grain stores and a gourmet supermarket of plump, grain-fed rodents awaited them. From this stage to the keeping and breeding of cats for increased vermin destruction must have been a simple step, since it benefited both sides.

With our efficient modern pest-control methods it is difficult for us to imagine the significance of the cat to those early civilizations, but a few facts about the attitudes of the ancient Egyptians towards their beloved felines will help to underline the importance that was placed upon them. They were, for instance, considered sacred, and the punishment for killing one was death. If a cat happened to die naturally in a house, all the human occupants had to enter full mourning, which included the shaving-off of their eyebrows.

Following death, the body of an Egyptian cat was embalmed with full ceremony, the corpse being bound in wrappings of different colours and its face covered with a sculptured wooden mask. Some were placed in cat-shaped wooden coffins, others were encased in plaited straws. They were buried in enormous feline cemeteries in huge numbers – literally millions of them.

The cat-goddess was called Bastet, meaning She-of-Bast. Bast was the city where the main cat temple was situated, and where each spring as many as half a million people converged for the sacred festival. About 100,000 mummified cats were buried at each of these festivals to honour the feline virgin-goddess (who was presumably a forerunner of the Virgin Mary). These Bastet festivals were said to be the most popular and best attended in the whole of ancient Egypt, a success perhaps not unconnected with the fact that they included wild orgiastic celebrations and 'ritual frenzies'. Indeed, the cult of the cat was so popular that it lasted for nearly 2,000 years. It was officially banned in AD 390, but by then it was already in serious decline. In its heyday, however, it reflected the immense esteem in which the cat was held in that ancient civilization, and the many beautiful bronze statues of cats that have survived bear testimony to the Egyptians' appreciation of its graceful form.

A sad contrast to the ancient worship of the cat is the vandalizing of the cat cemeteries by the British in the last century. One example will suffice: a consignment of 300,000 mummified cats was shipped to Liverpool where they were ground up for use as fertilizer on the fields of local farmers. All that survives from this episode is a single cat skull which is now in the British Museum.

The early Egyptians would probably have demanded 300,000 deaths for such sacrilege, having once torn a Roman soldier limb from limb for hurting a cat. They not only worshipped their cats, but also expressly prohibited their export. This led to repeated attempts to smuggle them out of the country as high-status house pets. The Phoenicians, who were the ancient equivalent of second-hand car salesmen, saw catnapping as an intriguing challenge and were soon shipping out high-priced moggies to the jaded rich all around the Mediterranean. This may have annoyed the Egyptians, but it was good news for the cat in those early days, because it introduced them to new areas as precious objects to be well treated.

Plagues of rodents that were sweeping Europe gave the cat a new boost as a pest-controller, and it rapidly spread across the continent. The Romans were largely responsible for this, and it was they who brought the cat to Britain. We know that cats were well treated in the centuries that followed because of the punishments that are recorded for killing one. These were not as extreme as in ancient Egypt, but fines of a lamb or a sheep were far from trivial. The penalty devised by one Welsh king in the tenth century reflected the significance to him of the dead cat. The animal's corpse was suspended by its tail with its nose touching the ground, and the punishment for its killer was to heap grain over the body until it disappeared beneath the mound. The confiscation of this grain gave a clear picture of how much a working cat was estimated to save from the bellies of rats and mice.

These good times for cats were not to last, however. In the Middle Ages the feline population of Europe was to experience several centuries of torture, torment and death at the instigation of the Christian Church. Because they had been involved in earlier pagan rituals, cats were proclaimed evil creatures, the agents of Satan and familiars of witches. Christians everywhere were urged to inflict as much pain and suffering on them as possible. The sacred had become the damned. Cats were publicly burned alive on Christian feast days. Hundreds of thousands of them were flayed, crucified, beaten, roasted and thrown from the tops of church towers at the urging of the priesthood, as part of a vicious purge against the supposed enemies of Christ.

Happily, the only legacy we have today of that miserable period in the history of the domestic cat is the surviving superstition that a black cat is connected with luck. The connection is not always clear, however, because as you travel from country to country the luck changes from good to bad, causing much confusion. In Britain, for instance, a black cat means good luck, whereas in America and continental Europe it usually means bad luck. In some regions this superstitious attitude is still taken remarkably seriously. For example, a few years ago a wealthy restaurant-owner was driving to his home south of Naples late one night when a black cat ran across the road in front of his car. He stopped and pulled in to the side of the road, unable to proceed unless the cat returned (to 'undo' the bad luck). Seeing him parked there on the lonely road late at night, a cruising police car pulled up and the officers questioned him. When they learned the reason, such was the strength of cat superstition that, refusing to drive on for fear of bringing bad luck on themselves, they also had to sit in their car and wait for the cat to return.

Although these superstitions still survive, the cat is now once again the much-loved house pet that it was in ancient Egypt. It may not be sacred, but it is greatly revered. The Church's cruel persecution has long since been rejected by ordinary people and, during the nineteenth century, a new phase of cat promotion exploded in the shape of competitive cat shows and pedigree cat breeding.

As already mentioned, the cat had not been bred into many different forms for different work tasks, like the dog, but there had been a number of local developments, with variations in colour, pattern and coat length arising almost accidentally in different countries. Travellers in the nineteenth century started to collect the strange-looking cats they met abroad and to transport them back to Victorian England. There they bred them carefully to intensify their special characteristics. Cat shows became increasingly popular, and during the past 150 years more than 100 different pedigree breeds have been standardized and registered in Europe and North America.

All these modern breeds appear to belong to one species: *Felis sylvestris*, the Wild Cat, and are capable of interbreeding, both with one another and with all races of the wild *sylvestris*. At the very start of feline domestication, the Egyptians began by taming the North African race of *Felis sylvestris*. Until recently this was thought of as a distinct species and was called *Felis lybica*. It is now known to be no more than a race and is designated as *Felis sylvestris lybica*. It is smaller and more slender than the European race of Wild Cat and was apparently easier to tame. But when the Romans progressed through Europe, bringing their domestic cats with them, some of the animals mated with the stockier northern Wild Cats and produced heavier, more robust offspring. Today's modern cats reflect this, some being big and sturdy, like many of the tabbies, while others are more elongated and angular, like the various Siamese breeds. It is likely that these Siamese animals and the other more slender breeds are closer to the Egyptian original, their domestic ancestors having been dispersed throughout the world without any contact with the heavy-set northern Wild Cats.

Although opinions have differed, it now seems highly improbable that any other species of wild feline was involved in the history of modern domestic cats. We do know that a second, bigger cat, *Felis chaus*, the Jungle Cat, was popular with the ancient Egyptians, but it appears to have dropped out of the running very early on. We can, however, be certain that it was originally a serious contender for the domestication stakes, because examination of mummified cats has revealed that some of them possessed the much larger Jungle Cat skulls. But although the Jungle Cat is one of the more friendly cats in captivity, it is huge in comparison with even the heftiest of modern domestic animals and it is therefore unlikely that it played a part in the later domestication story.

This is not the place to give details of the modern cat breeds, but a brief history of their introduction will help to give some idea of the way the modern cat 'fancy' has become established:

The oldest breeds involved are the various Shorthaired Cats, descendants of the animals spread across Europe by the Romans. There is then a long gap until the sixteenth century, when ships from the Orient arrived at the Isle of Man carrying a strange tail-less cat – the famous Manx. Because of its curiously mutilated appearance, this breed has never been widely popular, though it still has its devotees. At about the same time the first of the Longhaired Cats, the beautiful Angora, was brought into Europe from its Turkish homelands. Later, in the mid-nineteenth century, it was to be largely eclipsed by the even more spectacular Persian Cat from Asia Minor, with its enormously thick, luxuriant coat of fur.

Then, in the late nineteenth century, in complete contrast, the angular, elongated Siamese arrived from the Far East. With its unique personality – far more extrovert than other cats – it appealed to a quite different type of cat-owner. Whereas the Persian was the perfect, rounded, fluffy child substitute with a rather infantile, flattened face, the Siamese was a much more active companion.

At about the same time as the appearance of the Siamese, the elegant Russian Blue was imported from Russia and the tawny, wild-looking Abyssinian from what is now Ethiopia.

In the present century, the dusky Burmese was taken to the United States in the 1930s and from there to Europe. In the 1960s several unusual additions appeared as sudden mutations: the bizarre Sphynx, a naked cat from Canada; the crinkly-haired Rex from Devon and Cornwall; and the flattened-eared Fold Cat from Scotland. In the 1970s the Japanese Bobtail Cat, with its curious little stump making it look like a semi-Manx Cat, was imported into the United States; the crinkly Wire-haired Cat was developed from a mutation in America; and the diminutive Drain-Cat (so called because drains were a good place to hide in cat-scorning Singapore) appeared on the American scene, rejoicing in the exotic name of Singapura.

Finally there was the extraordinary Ragdoll Cat, with the strangest temperament of any feline. If picked up, it hangs limply like a rag doll. It is so placid that it gives the impression of being permanently drugged. Nothing seems to worry it. More of a hippie-cat than a hip-cat, it seems only appropriate that it was first bred in California.

This is by no means an exhaustive list, but it gives some idea of the range of cats available to the pedigree enthusiast. With many of the breeds I have mentioned there is a whole range of varieties and colour-types, dramatically increasing the list of show categories. Each time a new type of cat appears the fur flies – not from fighting felines, but from the unseemly skirmishes that break out between the overenthusiastic breed-ers of the new line and the unduly autocratic authorities that govern the major cat shows. Latest breed to top the tussle-charts is the aforementioned Ragdoll: ideal for invalids, say its defenders; too easy to injure, say its detractors.

To add to the complications, there are considerable disagree-ments between the different show authorities, with the Governing Council of the Cat Fancy in Britain recognizing different breeds from the Cat Fanciers' Association in America, and the two organizations sometimes confusingly giving different names to the same breed. None of this does much harm, however. It simply has the effect of adding the excitement of a great deal of heated argument and debate, while the pedigree cats themselves benefit from all the interest that is taken in them.

The seriousness with which competitive cat-showing is treated also helps to raise the status of all cats, so that the ordinary pet moggies benefit too in the long run. And they remain the vast majority of all modern domestic cats because, to most people, as Gertrude Stein might have said, a cat is a cat is a cat. The differences, fascinating though they are, remain very superficial. Every single cat carries with it an ancient inheritance of amazing sensory capacities, wonderful vocal utterances and body language, skilful hunting actions, elaborate territorial and status displays, strangely complex sexual behaviour and devoted parental care. It is an animal full of surprises, as we shall see on the pages that follow.

Why does a cat purr?

The answer seems obvious enough. A purring cat is a contented cat. This surely must be true. But it is not. Repeated observation has revealed that cats in great pain, injured, in labour and even dying often purr loud and long. These can hardly be called contented cats. It is true, of course, that contented cats do also purr, but contentment is by no means the sole condition for purring. A more precise explanation, which fits all cases, is that purring signals a friendly social mood, and it can be given as a signal to, say, a vet from an injured cat indicating the *need* for friendship, or as a signal to an owner, saying thank you for friendship given.

Purring first occurs when kittens are only a week old and its primary use is when they are being suckled by their mother. It acts then as a signal to her that all is well and that the milk supply is successfully reaching its destination. She can lie there, listening to the grateful purrs, and know without looking up that nothing has gone amiss. She in turn purrs to her kittens as they feed, telling them that she too is in a relaxed, co-operative mood. The use of purring among adult cats (and between adult cats and humans) is almost certainly secondary and is derived from this primal parent-offspring context.

An important distinction between small cats, like our domestic species, and the big cats, like lions and tigers, is that the latter cannot purr properly. The tiger will greet you with a friendly 'one-way purr' – a sort of juddering splutter – but it cannot produce the two-way purr of the domestic cat, which makes its whirring noise not only with each outward breath (like the tiger), but also with each inward breath. The exhalation/inhalation rhythm of feline purring can be performed with the mouth firmly shut (or full of nipple), and may be continued for hours on end if the conditions are right. In this respect small cats are one up on their giant relatives, but big cats have another feature which compensates for it – they can roar, which is something small cats can never do.

How does a cat purr?

It comes as a surprise to most people to learn that the experts are still arguing over something as basic as how a cat purrs. It is even more unexpected to find that the two alternative explanations are completely different from one another. This is not a dispute over some small detail, but about the basic mechanism of purring. Here are the rival theories:

The *false vocal cord* theory sees purring originating in the cat's voice-box, or larynx. In addition to the ordinary vocal cords, the cat possesses a second pair of structures called vestibular folds, or false vocal cords. The presence of this second pair of cords is thought to be the secret behind the extraordinary purring mechanism that permits the animal to produce the soft, rumbling sound for minutes and even hours on end, without any effort and without opening the mouth.

This theory regards purring as little more than heavy breathing of the type humans sometimes indulge in when they are asleep – in other words, snoring. With every inhalation and exhalation, air passes over the false vocal cords and makes the rrrrrrr noise of the purr. To produce this characteristic sound, air has to be interrupted by the contraction of the laryngeal muscles about thirty times a second.

The *turbulent blood* theory says that the cat's voice-box has nothing whatever to do with purring. Instead, it is argued that when the cat's blood-flow through its main veins into the heart is increased, turbulence is created. This is greatest at the point where the main vein, carrying blood from the animal's body back into its heart, is constricted as it passes through the animal's chest. The swirling blood is thought to make the purring noise, the diaphram acting as an amplifier of the vibrations. The noise thus created is thought to be passed up the animal's windpipe and into the sinus cavities of the skull, where it resonates to produce the purring sound. Some authorities believe that it is the arching of the back of the purring cat that increases this blood turbulence to create the purring sound, while others see the increase as being more to do with emotional changes affecting the animal's blood flow.

To the non-specialist there seems little doubt as to which of these theories provides the correct explanation. The false vocal cord theory is the most obvious and the simplest. It explains the otherwise puzzling presence of the second, or false, pair of vocal cords. We know that the true vocal cords are the ones that produce the ordinary vocalizations of the cat, such as miaowing, yowling and screaming, and if the false cords do not produce purring, then their presence requires some other explanation.

The blood turbulence theory has the merit of ingenuity, but little else. Anyone who has listened to a cat purring, using a stethoscope, will agree that when strongly amplified in this way, the purring sound simply does not have the quality of resonating blood turbulence. It is far too vibratory. And the idea that postural or emotional changes could bring on purring also seems far-fetched. The purring animal is relaxed, and common sense demands that this is precisely the time when any blood turbulence would subside, not increase.

Finally, the false vocal cord theory is supported by the evidence of the simple action of placing your fingers gently on the throat of a purring cat – there is no doubt, then, that the purring sound is stemming from the region of the voice-box, and it is hard to see why any alternative theory should have been put forward. That it *has*, despite the fact that superficially it is so unlikely, should, however, keep us on our toes. In biology, the most obvious explanation is not always the correct one, and much 'obvious' evidence has been disproved by careful, modern research. Where purring cats are concerned we must continue to keep an open mind.

How sensitive is the cat's hearing?

The cat's ears are much more sensitive than those of its owners, which is why cats loathe noisy homes. Loud music, screaming and shouting are torture to the delicate hearing apparatus of the typical feline.

It is the specialized hunting behaviour of cats that has resulted in their improved hearing. Although dogs have a much greater acoustic range than humans, cats exceed even dogs in their ability to hear high-pitched sounds. This is because humans and dogs rely most on chasing and trapping their prey, whereas cats prefer to lurk in ambush and listen very carefully for the tiniest sound. If they are to succeed as stealthy hunters, they must be able to detect the most minute rustlings and squeaks, and must be able to distinguish precise direction and distance to pinpoint their intended victims. This requires much more sensitivity than we possess, and laboratory tests have confirmed that domestic cats do, indeed, possess a very fine tuning ability.

At the lower level of sounds, there is little difference between humans, dogs and cats – this is not where it counts, if you are a hunter of small rodents and birds. At the higher levels, humans in the prime of life can hear noises up to about 20,000 cycles per second. This sinks to around 12,000 cycles per second in humans of retirement age. Dogs can manage up to 35,000 to 40,000 cycles per second, so that they are able to detect sounds that we cannot. Cats, on the other hand, can hear sounds up to an astonishing 100,000 cycles per second. This corresponds well to the high pitch of mouse sounds, which can be emitted up to this same level. So no mouse is safe from the alert ears of the predatory cat.

Opinions differ as to how high the cat can hear at full sensitivity. Some authorities believe that a really delicate reaction is only possible for the domestic cat at up to 45,000 cycles per second. Above this, they feel, there is a much weaker response. Most experts agree, however, that the figure is nearer 60,000 to 65,000 cycles per second, which would be enough to hear most prey sounds.

This acoustic ability of pet cats explains why they sometimes appear to have supernatural powers. They hear and understand the ultrasonic sounds that precede a noisy activity and respond appropriately before we have even realized that something unusual is going to happen. And do not overlook the ability of your snoozing feline. Even while taking a catnap the animal's ears are in operation. If something exciting is detected the cat is awake and responding in a split second. Perhaps this is why it sleeps twice as long as we do, making up in length of slumber what it lacks in depth.

Sadly for old cats, this wonderful sensitivity does not last forever. By the age of about five years a cat starts to lose its range of hearing and when the animal has become elderly, in feline terms, it is often nearly deaf. This explains why old cats sometimes fall victim to fast cars. It is not that they are too slow to avoid being hit, but that they simply do not hear the cars speeding towards them.

Younger cats are not only good at hearing high-pitched squeaks, they are also brilliant at detecting precise direction. They can distinguish between two sounds which are only 18 inches apart at a distance of 60 feet; they can separate with ease two sounds that are coming from the same direction, but at different distances; and they can differentiate two sounds that have only a half-tone of difference between them. In some tests this last ability was shown to one-tenth of a tone. Human composers must envy the cat its ears. But it is sad for the ordinary cat owner to realize that there is a whole range of sounds that simply cannot be shared with a much-loved pet. It is one of the humbling lessons we learn from living with companion animals, namely that we are in so many ways their inferiors, even though as a species we have come to dominate the planet.

How many sounds does a cat make?

A blind musician claimed that he could detect a hundred different sounds made by cats. American research workers, studying many hours of tape recordings, also insist that the feline vocal repertoire is huge – the most complex of any animal except *Homo sapiens*. Why should this be, when the wild cat is not a particularly sociable animal? In nature, it is those species that live in highly organized groups that require the most complicated communication systems. So there seems to be something odd about the cat.

The explanation is that the domestic cat uses two vocabularies at once. In the wild it would have one set of sounds for the mother–offspring relationship and would then replace that with another set for adult life. In the human home, the tame cat retains its infant vocalizations right through into adulthood and improves on them in the process. In addition, it also has all the usual howls and wails of the adult feline world of sex and violence. Put together this makes a truly impressive vocabulary.

Another way in which the cat's sound system appears much more complex than that of other species is in the degree of variation that can be applied to a single type of call. For a bird, an alarm call may be just a whistle and always that whistle. If the bird becomes more and more agitated, it simply repeats the whistle more often and perhaps faster. But when a cat is upset it can produce a whole variety of miaowing calls that sound different enough from one another to be classified as distinct vocalizations.

The danger with accepting that cats have a hundred different sounds is that the whole subject becomes too complicated to make much sense. The way of avoiding this is not to list all the sounds, but rather to list the signals or messages that the cat is trying to convey to its companions. When this is done, the language of the cat becomes much more clear. Here are some of the main messages and the noises that transmit them:

1 'I am angry'

When cats fight there is a terrible din. Since Chaucer's day this has been referred to as CATERWAULING, but it is often recorded not as an aggressive sound but as a sexual one. This error is found even in the august Oxford English Dictionary, where to caterwaul has been defined as 'To make the noise proper to cats at rutting time'. In fact, the sound is heard at any time when two cats are fighting and may have nothing whatever to do with sexual encounters. Two spayed females disputing a territorial boundary can caterwaul as dramatically as any 'rutting toms'.

The reason why caterwauling is associated with rutting is because it is most common then, the scent of the female on heat attracting males from far and wide and bringing them uncomfortably close together, so that they feel unusually inclined to vent their feelings of hostility towards one another.

The caterwauling of an aggressive cat is a perfect example of a single signal that takes many different forms. Each of these forms, or intensities, can be given a separate name, but they all belong to the same basic message: 'Clear off, or I will attack you.' Because the vocalization is prolonged and because the mood of hostility rises and falls second by second, the strength of the sound increases and decreases as well. As it does so, instead of simply becoming louder or softer, it changes its whole quality. Because of this, people speak of cats GROWLING, SNARLING, GURGLING, WAILING and HOWLING, to give just a few of the names that have been attached to this belligerent kind of sound production. All of these are varying intensities of the same basic sound unit and should be treated as such.

2 'I am frightened'

When a cat is afraid, its normal reaction is to run away silently, or to hide. There is no point in making a lot of noise at such a moment. But if a cat is cornered and cannot run away, even though it wishes fervently to do so, it may then make a sound which transmits the message: 'I fear you, but do not push me too far, or I will turn on you despite my fear.' A cat in such a situation may perform a strange, throaty YOWLING noise. This indicates that although it is very scared it has not entirely lost its aggression. Pressed further, it will lash out. More commonly, it will perform the SPIT and HISS display, especially if its tormentor is a large dog or an aggressive human being. Viewed objectively, these are strange sounds to offer an attacker. They are not very loud and in terms of the volume of sound produced they are not particularly impressive. And yet they do seem to work remarkably well, putting even the largest dog into a suddenly more respectful mood. There is a special reason for the sounds. Many mammals have an inborn fear of poisonous snakes and the defensive noises of a cornered snake are, of course, the spit and hiss. So the cat is almost certainly performing a mimicry display, reacting like a snake in the hope of triggering off some deep-seated fear of a venomous bite in the assailant's mind.

3 'I am in pain'

The SCREAM of a cat in agony is unmistakable and is similar to the scream of many other animals when being badly hurt or terrorized. It is sometimes described as a SHRIEK or a SCREECH, according to its intensity, and it is an adult version of the SQUEAL of a distressed kitten. For the kitten, it acts as a vital signal to its mother to come to the rescue. For the adult, it no longer provokes much help, since there is little mutual aid between adult felines, but where pet cats are involved it can, of course, act as a useful signal to the cat's pseudo-parents – its human owners – who may then, like surrogate mother cats, come to its rescue.

A special context in which the pain signal is given is at the end of the mating act, when the tom-cat is withdrawing his barbed penis from the female. This tears her and causes her a sharp pang of pain, making her cry out. Although pain is the dominant feeling she experiences at that moment, it is a pain that simultaneously makes her angry with the male, and she swipes out at him. The combination of acute pain with sudden hostility produces a harsher version of the usual cry of pain – more of a screech than a scream. The male reads it loud and clear and quickly attempts to avoid the sharp claws that are lunging towards him.

4 'I want attention'

For cat owners this is the most familiar sound to emanate from their feline friends. The MIAOW says many things in many contexts, but it always has the same basic message, namely 'I require your immediate attention'. It originates as a MEWING sound in tiny kittens, letting their mother know that they need some kind of help or that they are in some sort of trouble. In wild cats it then more or less disappears as they become adult, but domestic cats remain mentally like kittens even when they are fully grown, and continue to 'talk' to their human owners like kittens communicating with their mothers. And they do more. They start to refine their miaows in a way that wild cats never seem to do. They take the infantile mewing and they modify it to each situation in which they wish to express a need for something. There are begging miaows and demanding miaows, complaining miaows and anxious miaows. There are soft, flat miaows to be let out of the house and pitiful, drawn-out miaows to be let in again when it starts raining. There are expectant miaows when tin-opener noises are detected and irritated miaows when some fixed routine has been disregarded. An alert owner will know each of these variants on the 'I want attention' signal and may become quite fluent in 'cat-ese' after a few years.

Brave attempts have been made by certain experts to write down the different miaows, in order to classify and standardize them. The results are often hilarious, as when we are asked to distinguish between the MHRHRNAAAAHOU, the MHRHRHRN-NAAAAHOOOUUU and the MHHNGAAHOU. The human alphabet simply cannot cope, and unless you know the sounds already (in which case there is no point in writing them down) there is no hope of learning them from these strange strings of letters. Furthermore, since the refinement of the different miaows as your pet cat grows older is a personal affair between you and the cat, it is not surprising that there is considerable variation from one animal to another. They all start with the basic feline sound, genetically inherited like all the other elements of their communication system, but the artificial nature of the adult cat/human owner relationship creates a special world in which new subtleties develop that go beyond the genetically-shared vocabulary. Feline individualism begins to assert itself, and anyone who has owned, say, a crossbred tabby, a Siamese, a Persian and an Abyssinian will know that not only different individuals but also different breeds have their own special vocal characteristics. It would require a feline Professor Higgins to unravel all the intricacies of the language of the modern domesticated cat.

5 'Come with me'

When a mother cat wants her kittens to come near to her or to follow her, she gives a soft little CHIRRUPING noise. She may also use it as a greeting when she has been away from the kittens for a while. Adult domestic cats employ this same signal, which has been accurately described as a 'rising trill', when they are greeting their human owners. At such moments they are reversing the usual relationship and treating the humans as their kittens rather than as their mothers. (They also do this when they bring their owners dead birds from the garden, an action normally done only as part of the food-training routine when mother cats show their kittens the kind of prey they must later attack.) Significantly, the greeting trill is normally done when they are on the move, usually when they have come in from outside and are about to move off towards the place where they expect food to be waiting. So, although it sounds like a greeting, it probably still has some of the 'come along, follow me' meaning in it.

6 'I am inoffensive'

This is the famous feline PURRING sound. It is hard for some owners to accept that this noise means 'I am inoffensive' rather than 'I am content', but the fact is that inoffensiveness is the only condition to explain all the different contexts in which the purr occurs. It is essentially a signal that says the cat is in a non-hostile mood, is friendly, submissive, reassuring, appeasing, or, of course, contented. It is observed in the following cat-to-cat situations:

a When kittens are sucking at the nipple, letting the mother know all is well.

b When the mother is lying with her kittens, reassuring them that all is well.

c When the mother approaches the nest where the kittens are hiding, letting them know that there is nothing to fear by her approach.

d When a young cat approaches an adult for play, letting it

know that it is in a relaxed mood and accepts its subordinate social position.

e When a dominant adult cat approaches a young cat in a
friendly way, reassuring it of its non-hostile intentions.

f When an inferior cat is approached by a dominant
enemy, as an attempt to appease the more powerful one
with its submissive, non-hostile signal.

g When a sick cat is approached by a dominant one, letting
it know that it is in a weak, non-hostile mood.

Again, attempts have been made to classify purrs into different
types, but they all carry the same basic message of friendliness.
The nearest human equivalent to purring is smiling, and we
also make the same mistake when we say that a smiling man is
a happy man. Just like purring, smiling can occur when we are
being reassuring, submissive, non-hostile, and appeasing, as
well as when we are blissfully happy. Like the purr, the key
message of the smile is 'I am NOT going to do you any harm'.
As such it is vitally important for 'softening' social relationships and de-stressing moments of close proximity.

7 'I want to sink my teeth into you'

There is a strange little CLICKING noise made by cats that are on
the prowl and have spotted a prey animal. They use a variant of
it, a sort of teeth-gnashing, when they see a bird through a
window. It is no more than the action of clashing their teeth
together as if sinking them into the neck of the prey, in the
specialized killing bite of the cat, but it has become a sound
signal that many observers have commented on. Some authors
have suggested that the clicks are made to alert other cats to the
presence of the prey. One even made the mistake of describing
it as a signal to 'alert others in the pack', as if cats are group-
hunting animals like wolves. The fact that cats are solitary
hunters and would certainly not wish to let other felines join
them in stalking a particular prey, makes this clicking noise
something of a mystery. The only possible explanation is that
it is employed by mother cats when out with their nearly fully-
grown kittens, to help focus their attention on potential prey,
as part of a general hunt-training process.

These are the seven most important sound messages made by

domestic cats. With their many variants and subdivisions they provide a wonderfully expressive earful for the alert feline as it goes about its business. For human cat owners they offer a fascinatingly rich field of study and access to the world of their feline companions. Only occasionally do they jar on the nerves, as when your queen on heat starts her pitiful calling ('I want attention' – of a special kind) and attracts a chorus of savagely competitive, caterwauling toms to the space beneath your bedroom window. But the next time this happens, instead of hurling buckets of water, switch on a tape recorder and then, the next day, sit down and enjoy the feline opera at your leisure. If your sleep is not at risk, you will find the amazingly complicated sequence of moans, growls and yowls an absorbing example of prolonged animal communication.

How do cats react to music?

The correct answer appears to be idiosyncratically. Some cats show no interest whatever, while others detest it and still others adore it. It is hard, at first, to make any sense out of the reports that exist in the feline literature.

The French writer Theophile Gautier, for example, observed that his female cat would always listen attentively to the singers that he accompanied on the piano. She was not happy, however, when high notes were struck. They probably reminded her too closely of sounds of feline distress and she did her best to silence them. Whenever a female singer reached a high A, the cat would reach out and close the songstress's mouth with her paw. There must have been something specially feline about that particular note, because Gautier carried out experiments to see if he could fool the cat, but she always responded with her critical paw precisely when the note reached high A.

A more severe music critic was one of the cats owned by Frenchman C.C. Pierquin de Gembloux. This animal's reaction to certain sequences of notes was to throw itself into uncontrollable convulsions. A second cat, present at the same time, responded in a totally different way. Instead of having a fit, it jumped up and sat on the piano, listening to the same music with great interest.

The composer Henri Sauguet was astonished to discover that his cat, Cody, became ecstatic when it heard Debussy being played on the piano. It would roll around on the carpet, then leap on to the piano and then on to the pianist's lap, where it would start licking the hands that played the magic notes. When these same hands gave up playing under the onslaught of feline affection, the cat would wander off, but if they then began to play again, the cat immediately dashed back and resumed its licking.

Back in the 1930s two doctors by the names of Morin and Bachrach discovered to their surprise that the note of E of the fourth octave had the effect of making young cats defecate and adult ones become sexually excited. It was also noted that extremely high notes caused agitation in many cats.

What are we to deduce from these remarkable observations? Why on earth should cats have any reactions at all to something as sophisticated as human music? The answer seems to lie with the special signals that are given in feline 'language' by certain specific sounds. The mewing of a distressed kitten, for instance, is at a particular pitch, and if a musical note hits that pitch then it will disturb an adult cat, especially a female one. This may explain why Gautier's cat touched the mouths of the singers with her paw when they hit the particular note. She must have thought, at that moment, that each singer was a 'kitten' in distress, and she was no doubt trying to help in her own way.

Similarly, Sauguet's cat probably thought that his owner needed help, and rushed over to lick the hands from which the sounds seemed to be coming, as a way of comforting him – just as a mother cat would rush across to lick the fur of one of its kittens if the young animal appeared to be in distress.

The convulsions and sexual excitement of other cats are probably no more than erotic responses to sounds that remind felines of the courtship tones of their species. And fear induced by very high-pitched musical notes could simply be the natural panic reaction to what the cat hears as squeals of pain.

In other words, the musical sense of cats is just another feline myth. All they are doing – and with some remarkable individual variation – is responding to selected notes from the great array that music offers them, according to their own instinctive system of sound signals. Some musical notes trigger off parental feelings, others sexual ones and still others self-protection. The cats are mistaking our messages and we, in the past, have reciprocated by misunderstanding theirs.

Why do cats interfere when their owners answer the telephone?

Some cat owners report that whenever they are talking on the telephone their pet cats stalk over, leap up on the telephone table, and start rubbing against them, making it difficult sometimes to continue with the conversation. Gently removed to the floor, they are soon back again, as if jealous of the communication that is taking place between the owners and the voices at the other end of the line. Why do they act in this way?

The interpretation of jealousy is understandable, but incorrect. Although the owner feels that the cat's actions are a deliberate attempt to intervene in the conversations, the truth is that the cat is quite unaware of the person at the other end of the telephone line. All it observes is that, suddenly, its much-loved owner is talking. Furthermore, if there is nobody else in the room, it is clear to the cat that its owner must be talking to *it*. Most owners, it has already learned, do not talk to themselves, so there can be no other conclusions. So it responds appropriately, and moves in close to make its friendly response.

A closer scrutiny of such cases reveals that not all telephone calls get this treatment. If there is another human being in the room at the same time, the cat is less interested in the telephoner because it has long ago discovered that owners talk to their human friends and that this has no significance for cats. Also, the solitary telephoner may be less interesting when making boring calls than when making friendly ones. This is because cats become extremely sensitive to vocal tonality. If we speak in soft, loving tones on the telephone – the same tones we use when greeting our feline companions – then such calls will prove almost irresistibly attractive to cats in earshot. Who can blame them for returning our apparent greeting with the rubbings and nuzzlings of feline friendship?

How does a cat manage to fall on its feet?

Although cats are excellent climbers they do occasionally fall, and when this happens a special 'righting reflex' goes into instant operation. Without this a cat could easily break its back.

As it starts to fall, with its body upside-down, an automatic twisting reaction begins at the head end of the body. The head rotates first, until it is upright, then the front legs are brought up close to the face, ready to protect it from impact. (A blow to a cat's chin from underneath can be particularly serious.) Next, the upper part of the spine is twisted, bringing the front half of the body round in line with the head. Finally, the hind legs are bent up, so that all four limbs are now ready for touchdown and, as this happens, the cat twists the rear half of its body round to catch up with the front. Finally, as it is about to make contact, it stretches all four legs out towards the ground and arches its back, as a way of reducing the force of the impact.

While this body-twisting is taking place, the stiffened tail is rotating like a propeller, acting as a counterbalancing device. All this occurs in a fraction of a second and it requires slow-motion film to analyse these rapid stages of the righting response.

How does a cat use its whiskers?

The usual answer is that the whiskers are feelers that enable a cat to tell whether a gap is wide enough for it to squeeze through, but the truth is more complicated and more remarkable. In addition to their obvious role as feelers sensitive to touch, the whiskers also operate as air-current detectors. As the cat moves along in the dark it needs to manoeuvre past solid objects without touching them. Each solid object it approaches causes slight eddies in the air, minute disturbances in the currents of air movement, and the cat's whiskers are so amazingly sensitive that they can read these air changes and respond to the presence of solid obstacles even without touching them.

The whiskers are especially important – indeed vital – when the cat hunts at night. We know this from the following observations: a cat with perfect whiskers can kill cleanly both in the light and in the dark. A cat with damaged whiskers can kill cleanly only in the light; in the dark it misjudges its killing-bite and plunges its teeth into the wrong part of the prey's body. This means that in the dark, where accurate vision is impeded, healthy whiskers are capable of acting as a highly sensitive guidance system. They have an astonishing, split-second ability to check the body outline of the victim and direct the cat's bite to the nape of the unfortunate animal's neck. Somehow the tips of the whiskers must read off the details of the shape of the prey, like a blind man reading braille, and in an instant tell the cat how to react. Photographs of cats carrying mice in their jaws after catching them reveal that the whiskers are almost wrapped around the rodent's body, continuing to transmit information about the slightest movement, should the prey still be alive. Since the cat is by nature predominantly a nocturnal hunter, its whiskers are clearly crucial to its survival.

Anatomically the whiskers are greatly enlarged and stiffened hairs more than twice the thickness of ordinary hairs. They are embedded in the tissue of the cat's upper lip to a depth three times that of other hairs, and they are supplied with a mass of nerve-endings which transmit the information about any contact they make or any changes in air-pressure. On average the cat has twenty-four whiskers, twelve on each side of the nose, arranged in four horizontal rows. They are capable of moving both forwards, when the cat is inquisitive, threatening, or testing something, and backwards, when it is defensive or deliberately avoiding touching something. The top two rows can be moved independently of the bottom two, and the strongest whiskers are in rows two and three.

Technically whiskers are called vibrissae and the cat has a number of these reinforced hairs on other parts of its body – a few on the cheeks, over the eyes, on the chin and, surprisingly, at the backs of the front legs. All are sensitive detectors of movement, but it is the excessively long whiskers that are by far the most important vibrissae, and it is entirely apt that when we say that something is 'the cat's whiskers' we mean that it is rather special.

Why do cats' eyes glow in the dark?

Because they possess an image-intensifying device at the rear of their eyes. This is a light-reflecting layer called the *tapetum lucidum* (meaning literally 'bright carpet'), which acts rather like a mirror behind the retina, reflecting light back to the retinal cells. With this, the cat can utilize every scrap of light that enters its eyes. With our eyes we absorb far less of the light which enters them. Because of this difference cats can make out movements and objects in the semi-darkness which would be quite invisible to us.

Despite this efficient nocturnal ability it is not true that cats can see in complete darkness, as some people seem to believe. On a pitch black night they must navigate by sound, smell and the sensitivity of their amazing whiskers.

Why do cats' eyes contract to a vertical slit?

Reducing the pupils to slits, rather than tiny circles, gives the cat a more refined control over precisely how much light enters the eyes. For an animal with eyes sensitive enough to see in very dim light it is important not to be dazzled by bright sunlight, and the narrowing of the pupils to tight slits gives a greater and more accurate ability to cut down the light input. The reason why cats have vertical slits rather than horizontal ones is that they can use the closing of the lids to reduce the light input even further. With these two slits – the vertical one of the pupil and the horizontal one of the eyelids – working at right angles to one another, the feline eye has the possibility of making the most delicate adjustment of any animal, when faced with what would otherwise be a blinding light.

Confirmation of the fact that it is the nocturnal sensitivity of the cat's eyes that is linked with the contraction of the pupils to slits, is found in the observation that lions, which are daytime killers, have eyes that contract, like ours, to circular pinpricks.

Can cats see colours?

Yes, but rather poorly, is the answer. In the first half of this century scientists were convinced that cats were totally colour-blind and one authority reworked a popular saying with the words: 'Day and night, all cats see grey.' That was the prevailing attitude in the 1940s, but during the past few decades more careful research has been carried out and it is now known that cats can distinguish between certain colours, but not, apparently, with much finesse.

The reason why earlier experiments failed to reveal the existence of feline colour vision was because in discrimination tests cats quickly latched on to subtle differences in the degree of greyness of colours and then refused to abandon these clues when they were presented with two colours of exactly the same degree of greyness. So the tests gave negative results. Using more sophisticated methods, recent studies have been able to prove that cats can distinguish between red and green, red and blue, red and grey, green and blue, green and grey, blue and grey, yellow and blue, and yellow and grey. Whether they can distinguish between other pairs of colours is still in dispute. For example, one authority believes that they can also tell the difference between red and yellow, but another does not.

Whatever the final results of these investigations one thing is certain: colour is not as important in the lives of cats as it is in our own lives. Their eyes are much more attuned to seeing in dim light, where they need only one-sixth of the light we do to make out the same details of movement and shape.

What does a cat signal with its eyes?

The next time you feed your cat, take a close look at its eyes. If it is hungry, the moment the food dish appears the animal's pupils will dilate. The vertical slits will expand to dark pools of feline expectation. Careful tests have shown that when this happens the area of the pupils may increase to between four and five times their previous size in less than one second.

This dramatic change is part of the cat's mood-signalling system, but it is only one way in which the eyes change their expression. The most basic eye change is connected with variations in light intensity. The more light that falls on the eyes, the more the pupils contract to vertical slits; the less light there is, the more they open up to round, black pools. This type of alteration in the appearance of the eye goes on all day, as the animal moves from light to shade and back again, and it is so common a shift that it tends to obscure the other pupil changes that are taking place.

Proximity of the viewed scene also affects the cat's pupils: the closer an object is, the more it has to constrict them; with more distant objects it expands them a little. This type of change also interferes with our reading of the mood-signals coming from the eyes.

To make matters even more complicated, there are two quite distinct kinds of mood-change, for the cat's pupils will become greatly enlarged not only when it sees something pleasant but also when it sees something terribly threatening. The only way to clarify this state of affairs is to say that if you see a cat's pupils suddenly expand, without any change in light intensity or proximity of objects, then it is experiencing a state of strong emotional arousal. This arousal may be pleasant, as with the arrival of tasty food, or unpleasant, as with the arrival of a large aggressive rival. In both cases, the pupils will enlarge more than normal, as if trying to increase the input of information from the exciting stimuli.

Because a frightened, defensive cat shows extreme pupil

expansion, an opposite signal seems to have evolved for its dominant, aggressive and fearless rival. For such an animal there is only one possible kind of eye expression: the narrow vertical slit of a fully contracted pupil. But take care! This does not mean that only a slit-eyed cat is dangerous. A frightened cat with expanded pupils is just as likely to strike out in panic. In fact, a submissive cat in the house that has 'had enough' and is about to defend itself will often dilate its pupils rapidly just before it lashes out at you. So it is important to read the 'expanded pupil' signal very carefully and to place it in context before interpreting it.

In addition to pupil changes there is also the possibility of signalling mood by the degree of opening or closing the eyelids. An alert cat has fully-opened eyes and this is the condition that is always maintained in the presence of strangers, who are not entirely trusted by the cat. If the animal switches to half-closed eyes, this is an expression of total relaxation signalling complete trust in the friendship of its owners.

Full closure of the eyes only occurs in two contexts: sleep and appeasement. When two cats are fighting and one is forced into submission, it often performs what is called 'cut-off', where it turns away from its tormentor and shuts its eyes, trying to blot out the frightening image of its dominant rival. This is basically a protective action, an attempt to save the eyes from possible danger, but it has also become a way of reducing the unbearable tension of the moment. In addition, the victor sees it as a sign of capitulation by his opponent.

Finally a word about the stare. Prolonged staring with wide-open eyes has a special significance for the cat. It is an eye signal that denotes aggression. In other words, for a human being to stare at a cat is to threaten it. Yet we are always doing this, because we so like to enjoy contemplating the beauty of our feline companions. We do it innocently, without wishing the cat any harm, but it is sometimes difficult for the animal to appreciate this. The solution is to enjoy staring at the cat at moments when it is not staring back. If we lock eyes with it we are unavoidably intimidating it, when this is the last thing we wish to do. By a small adjustment, however, we can greatly improve our relationship and make the cat feel much more comfortable and at ease in our presence.

How does the cat use its haws?

The haw is the cat's third eyelid. It is situated at the inner corner of the eye and comes into action to protect the delicate organ from damage or to lubricate the corneal surface by spreading the cat's tears evenly across it. When the haws are activated, they move sideways across the eye and then return to their resting position. In this respect the cat has an advantage over the human species, for we are unable to move our third eyelids, which exist only as small pink lumps at the inner point of each eye.

The cat's haw – or nictitating membrane, to give it its official name – is not normally conspicuous, but if the cat is in ill health, undernourished, or about to succumb to a major disease, it may become permanently visible, giving the cat's eye a 'half-shuttered' look. When this happens, it is an important clue that the animal is in need of veterinary assistance. The appearance of the haws in these circumstances is caused by the fact that there are shock absorber pads of fat behind the eyeballs that start to shrink if the animal's health is below par. This shrinkage means that the eyes sink into the head slightly, and this in turn causes the haws to move forward and half cover the corneal surfaces. When the cat returns to full health again, the fat pads are replenished and the eyes pushed forward again, hiding the haws once more.

Why does a cat like being stroked?

Because it looks upon humans as 'mother cats'. Kittens are repeatedly licked by their mothers during their earliest days and the action of human stroking has much the same feel on the fur as feline licking. To the kitten, the mother cat is 'the one who feeds, cleans and protects'. Because humans continue to do this for their pets long after their kitten days are behind them, the domesticated animals never fully grow up. They may become full-sized and sexually mature, but in their minds they remain kittens in relation to their human owners.

For this reason cats – even elderly cats – keep on begging for maternal attention from their owners, pushing up to them and gazing at them longingly, waiting for the pseudo-maternal hand to start acting like a giant tongue again, smoothing and tugging at their fur. One very characteristic body action they perform when they are being stroked, as they greet their 'mothers', is the stiff erection of their tails. This is typical of young kittens receiving attention from their real mothers and it is an invitation to her to examine their anal regions.

Why does a cat roll over to lie on its back when it sees you?

When you enter a room where a cat is lying asleep on the floor and you greet it with a few friendly words, it may respond by rolling over on its back, stretching out its legs as far as they will go, yawning, exercising its claws and gently twitching the tip of its tail. As it performs these actions, it stares at you, checking your mood. This is a cat's way of offering you a passively friendly reaction and it is something which is only done to close family intimates. Few cats would risk such a greeting if the person entering the room were a stranger, because the belly-up posture makes the animal highly vulnerable. Indeed, this is the essence of its friendliness. The cat is saying, in effect, 'I roll over to show you my belly to demonstrate that I trust you enough to adopt this highly vulnerable posture in your presence.'

A more active cat would rush over to you and start rubbing against you as a form of friendly greeting, but a cat in a lazy, sleepy mood prefers the belly-roll display. The yawning and stretching that accompany it reflect the sleepiness of the animal – a sleepiness which it is prepared to interrupt just so much and no more. The slight twitching of the tail indicates that there is a tiny element of conflict developing – a conflict between remaining stretched out and jumping up to approach the new arrival.

It is not always safe to assume that a cat making this belly-up display is prepared to allow you to stroke its soft underside. It may appear to be offering this option, but frequently an attempt to respond with a friendly hand is met with a swipe from an irritated paw. The belly region is so well protected by the cat that it finds contact there unpleasant, except in relationships where the cat and its human owner have developed a very high degree of social intimacy. Such a cat may trust its human family to do almost anything to it. But the more typical, wary cat draws the line at approaches to its softer parts.

Why does a cat rub up against your leg when it greets you?

Partly to make friendly physical contact with you, but there is more to it than that. The cat usually starts by pressing against you with the top of its head or the side of its face, then rubs all along its flank and finally may slightly twine its tail around you. After this it looks up and then repeats the process, sometimes several times. If you reach down and stroke the animal, it increases its rubbing, often pushing the side of its mouth against your hand, or nudging upwards with the top of its head. Then eventually it wanders off, its greeting ritual complete, sits down and washes its flank fur.

All these elements have special meanings. Essentially what the cat is doing is implementing a scent-exchange between you and it. There are special scent glands on the temples and at the gape of the mouth. Another is situated at the root of the tail. Without your realizing it, your cat has marked you with its scent from these glands. The feline fragrances are too delicate for our crude noses, but it is important that friendly members of the cat's family should be scent-sharing in this way. This makes the cat feel more at home with its human companions. And it is important, too, for the cat to read *our* scent signals. This is achieved by the flank-rubbing element of the greeting, followed by the cat sitting down and 'tasting' us with its tongue – through the simple process of licking the fur it has just rubbed so carefully against us.

Why do some cats hop up on their hind legs when greeting you?

One of the problems cats have when adjusting to human companions is that we are much too tall for them. They hear our voices coming from what is, to them, a great height and they find it hard to greet such a giant in the usual way. How can they perform the typical cat-to-cat greeting of rubbing faces with one another? The answer is that they cannot. They have to make do with rubbing our legs or a downstretched hand. But it is in their nature to aim their greetings more towards the head region, and so they make a little intention movement of doing this – the stiff-legged hop in which the front feet are lifted up off the ground together, raising the body for a brief moment before letting it fall back again to its usual four-footed posture. This greeting hop is therefore a token survival of a head-to-head contact.

A clue to this interpretation comes from the way small kittens sometimes greet their mother when she returns to the nest. If they have developed to the point where their legs are strong enough for the 'hop', the kittens will peform a modest version of the same movement, as they push their heads up towards that of the mother cat. In their case there is not far to go, and she helps by lowering her own head towards theirs, but the incipient hop is clear enough.

As with all rubbing-greetings, the head-to-head contact is a feline method of mingling personal scents and turning them into shared family scents. Some cats use their initiative to re-create a better head contact when greeting their human friends. Instead of the rather sad little symbolic hop, they leap up on to a piece of furniture near the human and employ this elevated position to get themselves closer for a more effective face-to-face rub.

Why does a cat trample on your lap with its front paws?

All cat-owners have experienced the moment when their cat jumps up and with cautious movements settles itself down on their lap. After a short pause it starts to press down, first with one front paw and then with the other, alternating them in a rhythmic kneading or trampling action. The rhythm is slow and deliberate as if the animal is marking time in slow motion. As the action becomes more intense the prick of claws can be felt, and at this point the owner usually becomes irritated and shoos the cat away, or gently picks it up and places it on the floor. The cat is clearly upset by this rebuff and the owners are similarly put out when, brushing away a few cat hairs, they discover that the animal has been dribbling while trampling. What does all this mean?

To find the answer it is necessary to watch kittens feeding at the nipple. There the same actions can be observed, with the kittens' tiny paws kneading away at their mother's belly. These are the movements which stimulate the flow of milk to the nipples and the dribbling is part of the mouthwatering anticipation of delicious nourishment to come. This 'milk-treading', as it is called, is done at a very slow pace of approximately one stroke every two seconds, and it is always accompanied by loud purring. What happens when the adult pet tramples on the lap of its human owner must therefore be interpreted as a piece of infantile behaviour. It would seem that when the owner sits down in a relaxed manner, signals are given off saying to the cat, 'I am your mother lying down ready to feed you at the breast.' The adult cat then proceeds to revert to kittenhood and squats there, purring contentedly and going through the motions of stimulating a milk supply.

From the cat's point of view this is a warm, loving moment and its bodily removal by a claw-pricked owner must be quite inexplicable. No good cat mother would behave in such a negative way. People react rather differently. To the cat they are clearly maternal figures, because they do supply milk (in a saucer) and other nourishment, and they do sit down showing their undersides in an inviting manner, but once the juvenile reaction of milk-treading is given, they suddenly and mystifyingly become upset and thrust the pseudo-infant from them.

This is a classic example of the way in which interactions between humans and cats can lead to misunderstandings. Many can be avoided by recognizing the fact that an adult domestic cat remains a kitten in its behaviour towards its pseudo-parental owner.

Why does a cat sometimes bite the hand that strokes it?

Some cats are completely docile and allow as much petting and stroking as their human companions wish to lavish upon them. Others, if they have had enough attention, will simply start to struggle and then leap down or move away. But there is another type of cat – more common than most people suppose – that has a more violent reaction to being over-petted. Described by one author as the Jekyll and Hyde Cat, this animal suddenly lashes out and attacks its friendly owner's hand. The assault is so unexpected and apparently unjustified that it leaves the owner not only bleeding but also deeply perplexed.

Before explaining the cause of this reaction, it helps to observe precisely what happens. First the owner starts to stroke the cat, tickle its ear, or gently rub its head. The cat responds lovingly, totally relaxed and probably purring. Then, after a while, it stiffens imperceptibly and, often unnoticed by the stroker, its ears start to rotate so that their backs face forward. This is the key danger signal. At the same moment the pupils may dilate. Then, so fast that the movements are almost impossible to analyze, the cat lashes out with its claws extended, raking the skin of the hand in front of it. At the same time it may make a sudden, savage bite with its canines. Then in a flash it dashes away as if fleeing in panic.

Essentially this is the behaviour of a cat that feels itself severely threatened and strikes out to protect itself. Having done this and expecting immediate reprisals, it then runs for cover. But why does it suddenly feel threatened by the contact of its peaceful owner? There appear to be two possible explanations. The first has to do with the individual animal's past history. It often happens that a particular cat finds itself betrayed by a friendly human hand. The fingers move gently in and start to tickle it or rub its face and then, without warning, grab it and pick it up. This is the strategy employed by strangers who are fearful of the cat's claws and have to find some way of picking it up without being attacked. A vet who wishes to examine a cat may approach it in this way before, for instance, holding it down to give it an injection. Cats have long memories, especially where a nasty shock is concerned, and may remember a bad experience of this kind for years afterwards. This creates a conflict for them because, although they want to be stroked and petted like any other domestic cat, they are deeply suspicious of the hand that does the petting, fearing that at any moment it may grab them and hold them down. At the start of the contact the need to be stroked is so strong that it suppresses their fear. As the seconds tick by, however, and this need becomes increasingly satisfied, the fear of being grabbed starts to well up inside them. Suddenly it takes over — uncontrollably, they lash out and then flee in panic, their ancient memory switching them from Jekyll to Hyde.

For cats without this unpleasant memory there is less chance that they will bite the hand that strokes them, but it can still happen occasionally. To understand this it is necessary to consider the meaning of the stroking and petting, not from our point of view but from the cat's.

Adult cats occasionally groom one another, but the most common form of social grooming is the licking of kittens by their mother. The kittens tolerate a certain amount of this before deciding that enough is enough. To the adult pet cat living with human companions, its owner's hand is a symbolic 'mother's tongue' tugging at or smoothing its fur. When it has had enough 'cleaning' its mood changes and the hand ceases to fill the role of a maternal tongue. Without changing its movements, it now becomes the 'giant paw' of a huge cat — in this new role it is suddenly threatening and the animal responds appropriately with a defensive reaction.

Why are cats attracted to people who dislike them?

If a cat enters a room where several people are talking, it is very likely to make for the one person there who has an abnormal fear of felines. To that individual's horrified disbelief, the animal then proceeds to rub around his or her legs and may even jump up on the person's lap. Why is the cat so perverse?

To some people, this is a confirmation of the old idea that there is something inherently wicked in the feline personality, and that the animal deliberately selects someone with a cat phobia and then sets out to cause them embarrassment. But this kind of superstitious romanticizing is superfluous. There is a much simpler explanation of the animal's behaviour.

When the cat enters the room and looks around, it notices that several people are staring at it. They are the cat-lovers, gazing at the cat because they like it. But, in feline terms, to be stared at is to be mildly threatened. Children are sometimes told 'it is rude to stare', but cat-lovers often forget this rule when they are looking at an approaching feline. Instead of a glance, which is always acceptable, they keep staring at the animal in a way that makes it feel uncomfortable. The only person there not doing this is the cat-hater, who looks away and keeps very still, trying to be ignored by the feared animal. But such behaviour has precisely the opposite effect. For the cat, in search of a friendly lap on which to sit, makes a bee-line for this ideal companion, who is *not* moving around, *not* waving hands about, *not* making shrill remarks and, above all, *not* staring. The cat is thus showing its appreciation of the non-intimidating body-language.

The secret for any cat-phobic individuals who want to keep their distance is to lean towards a cat, stare fixedly at it with wide-open eyes and make agitated hand movements, asking the cat in strident tones to come and sit on their laps. This will have exactly the opposite effect and they will then be able to relax without appearing to have insulted a host's favourite pet.

Why do some cats hate men?

Some owners notice that their cats love human females and hate or fear human males. Why do they make this distinction?

There are two factors at work here. One concerns the tonality of the human voice. The female voice, being much higher pitched, is closer in quality to that of the cat, which makes it more appealing. But this is not sufficient to explain the behaviour of some felines which will run and hide as soon as they hear a man approaching. In such cases it usually means that a man has, at some time in the past, hurt the cat in question. Cats have long memories and if they have suffered pain at male hands they may hate all males for months afterwards, sometimes literally for years.

This suggests that men are more cruel than women towards cats, but such an interpretation assumes that cats can distinguish between deliberate cruelty and pain inflicted for the animals' own good. The problem is that the majority of vets are men and cats often take a very long time to forget the helpful attentions they receive at their local animal clinics. The vet is only trying to assist the cat, but there is no way that the animal can associate the injection, or the forcible administration of a medicine, with its feeling healthier later on. All it knows is that when a human male came close to it, it was restrained and then something painful happened. So the next time a male voice, or a male footfall (for cats can quickly associate the heavy male tread with the deeper male voice), or the scent of a human male (for cats can also associate human odours with human genders) is detected, the cat thinks that trouble is on its way again, and beats a rapid retreat.

One particular cat, that was originally very friendly towards both men and women, demonstrated this extremely clearly. It was a stray that came to beg for food in the garden of a town house. When fed, it was happy to rub up against the legs of its new human friends, of either sex. Then someone offered to adopt the cat and the local vet was called in to give the animal a clean bill of health before it was moved to its new, permanent home. Cornered in a garden shed, the frightened cat was caught by the vet and given protective injections. It was then boxed up and taken to meet its new owner, an elderly woman living alone. The cat gradually made friends with her and became extremely loving, sleeping on her bed and following her everywhere, rubbing against her and jumping up on her lap. From a starving stray it was transformed into a contented house cat. But the memory of being caught and injected never faded, and even a year later the cat would still rush upstairs and hide beneath the bed if ever a man visited the house. In its mind it was convinced that the vet had returned to hurt it again. No amount of male kindness or pleading could reduce its fear. Visiting women, on the other hand, were always welcome and were treated to leg-rubbing and purring with as much friendship as was shown to its new owner. Its distinction between human males and human females was almost complete. There was only one exception, a very elderly male who sat quietly and hardly moved. Somehow the cat was capable of 'separating' in its mind this elderly man from the image of the much younger vet. It must be one of the major frustrations of being a veterinary surgeon that you can never convince your patients that you are on their side and are working tirelessly to save them and help them back to full health.

People who, with great kindness, acquire a rescued cat from an animal sanctuary, should be prepared for unusual reactions of this kind. They must never blame themselves for what happens, if the cat reacts fearfully to certain individuals or certain specific conditions. In such cases the animal is living in its past, a past about which the new owner knows little or nothing. Early traumas come back to haunt such cats and force them to behave in strange ways. Much patience is needed, but when eventually the cat does accept its new home, it will probably become the most loving and affectionate of all feline pets. It may even become over-affectionate, refusing to leave its rescuer alone, but this is simply a measure of its inner anxiety concerning the possibility that it may be betrayed once again. It is sometimes hard for cats to understand people and, like elephants, they never forget a shock encounter.

Why do cats sulk?

A scolded cat often turns its back on its owners and haughtily refuses to look at them. One owner describes this 'cold shoulder treatment' in the following words: 'He turns his back, sits down neatly and deliberately, and won't answer if we call his name as he usually does, though he sometimes puts one ear back.' This behaviour, observed by many owners when their pets have been chastised or corrected in some way, is usually referred to as a dignified sulk. But what is the cat really doing?

The answer is not that it is demonstrating 'wounded pride', as its owners believe, but instead is revealing its social inferiority. Its haughtiness is apparent, not real. This is hard for some owners to accept, because they have such respect for their feline companions. But they overlook the fact that, to the cat, they are huge and therefore psychologically overwhelming. When a cat misbehaves and its owner reacts crossly, the cat feels threatened. An owner's anger at some feline misdeed usually involves harsh tones and fixed staring. Staring is very intimidating to a cat and its natural response is to avoid the hostile image of the staring eyes. This it does by turning away in a deliberate manner and refusing to look again at the glowering face. Hence the apparent haughtiness of the 'turn-my-back-on-you' posture.

This action is called 'cut-off' because it cuts off the input — the hostile face looming over it. It has a double effect: it reduces the fear in the cat itself and enables it to stay where it is, rather than move off into the distance; it also prevents any counter-staring by the cat, which would spell defiance and possibly provoke further hostility.

The importance of this 'anti-stare' in feline social life is evident whenever two cats are involved in a status battle. The dominant cat always keeps a fixed stare directed towards its rival. The subordinate cat, if it wants to hold its ground, deliberately looks right away from its enemy and makes absolutely sure that its gaze never goes anywhere near the eyes of the glaring overlord. In the human context, this threat-stare has become a regular ritual of major boxing matches. When the referee talks to the two boxers just before the first round, each fighter stares closely and directly into the opponent's eyes. Neither dares to look away for an instant, in case this is read as a sign of weakness. For the 'sulking' cat, this sign of weakness is being deliberately displayed as a response to its owner's threats.

Any people who doubt this can carry out a simple test devised by the great cat authority, Paul Leyhausen, the next time they visit a zoo. Leyhausen proved the power of the direct stare by standing in front of a tiger's cage and hiding his eyes. He did this by bringing a camera up to his face, through which he could nevertheless still see the tiger's actions. The animal crouched ready for an attack and then dashed across its cage floor towards the spot where Leyhausen was standing. As it came near he quickly lowered the camera and directed a wide-eyed stare, straight at the big cat. It skidded to a halt immediately and rapidly looked away, avoiding the man's gaze. As soon as he covered his eyes with his camera again, another attack was launched. Again he froze it with a quick stare, and was able to repeat this process time and again.

Apart from the fact that this provides a valuable lesson for anyone unexpectedly encountering a big cat at close quarters, it explains the way in which lion-tamers at a circus manage to dominate their animal companions. A fixed stare from the trainer and they look away, remaining placidly ('sulkily') in their place.

The stare-threat phenomenon also explains another oddity of feline behaviour. Some observers have noticed that domestic cats, when hunting small birds in the garden, appear to be amazingly intelligent in one particular respect. If the bird's head disappears behind some small obstruction, the cat can be seen to rush forward and pounce, as though it knows that at that moment the bird cannot see its rapid advance. For the cat to reason this out would require considerable mental agility, but there is of course a simpler explanation. As long as the bird's eye is visible, it is automatically giving the cat a 'stare' that inhibits its attacking lunge. Once the eye is accidentally hidden behind some obstruction, the stare is switched off and the cat can attack. Studies of big cats stalking prey have revealed a similar interaction. If the prey looks up and stares straight at the lion or tiger, the big cat looks sheepishly away as if suddenly indifferent to the whole business of predation. So for any prey with the courage to hold its ground and out-stare a hunting lion, there is some considerable advantage to be gained . . . unless of course there is another lion coming up from behind . . .

Why do cats suddenly make mad dashes around the house?

Cat owners often notice that their pet will suddenly and for no apparent reason make a headlong dash through the house. Moving at top speed, the animal positively flings itself along and then, just as suddenly, comes to rest as though nothing strange has occurred. When this behaviour is seen for the first time, the owners may become seriously concerned, imagining that the cat is having some kind of fit or seizure. Unaware of how common this outburst of activity really is among housebound cats, they interpret it as some sort of abnormality and may even call in the vet to examine the animal. But their fears are groundless. There is nothing unusual about the feline 'mad dash'. Almost all cats do it and there is a simple explanation.

The mad dash is what is called a 'vacuum' activity. Cats kept indoors a great deal, with every whim catered to and with plenty of food always available, eventually come to suffer from a special kind of deprivation. They lack the opportunity to express their inborn urges to hunt and to flee from danger. There is no prey to catch and no predators or rivals from which to escape. Day after day, the finely-tuned responses that all felines possess – to make a sudden rush towards an unsuspecting mouse, or to make a headlong flight from approaching danger – are thwarted by the peace and luxurious calm of the home in which they live. They reach a point where even the smallest stimulus will trigger off a massive reaction. The pent-up energy overflows and a mad dash is on.

Proof that these outbursts are overflow or vacuum activities rather than pathological fits can be found by comparing the indoor behaviour of hard-living rural cats, with that of lap-of-luxury town cats. The working cats spend much of their time outside chasing birds and rodents, or challenging rival cats, and are wonderfully relaxed when they do, at last, come indoors for a saucer of milk or a snooze by the fire. Their main preoccupation is licking and grooming themselves and their most vigorous activity is likely to be the languorous stretching of tired limbs. The pampered town cat, by contrast, is often seen to be prowling around the house like a caged tiger. Even if it does go outside, there is little to hunt on the manicured lawns, and no serious threat from feline neighbours. So, when it returns for yet another pre-killed dish of food, it may find itself in a listless, frustrated mood. After resting for a while, it suddenly gets up, looks around and then sets off on one of its mad dashes. In this way it is able to release some of its hunting or fleeing energy and to feel more relaxed again.

Some owners report that their cats are liable to do this immediately after using the litter box. Others see it occurring at exactly the same time each day, but with no obvious connection with any other activity. Still others observe that it usually starts with a 'trigger action' of some sort. Such actions are usually in the form of a mock attack from a human friend, the presence of some kind of 'ghost prey', or some sharp, sudden sound or movement. The cats use these triggers as an 'excuse', so to speak, to release the frustrated response. Sometimes they even seem to provoke the trigger actions deliberately. A cat may approach its human owner and purposely make a nuisance of itself in a way that it has learnt will cause anger. When the owner shouts at the cat, instead of simply stopping as it usually does, the cat will massively over-react with one of its mad-panic rushes. The response is out of all proportion to the stimulus, and this is what makes it special, distinguishing it from occasions where a cat merely withdraws after being scolded.

How delicate is the cat's sense of taste?

Since cats can see, hear and smell with more sensitivity than we can, it is reassuring to find that in one respect at least we have superior sense organs. When it comes to taste, our tongues are slightly better than theirs. But only slightly. Like us, cats are responsive to four basic tastes – sour, bitter, salt and sweet. We respond to all four strongly, but cats are weak when it comes to sweet tastes. They lack our 'sweet tooth'.

Until recently many authorities stated categorically that cats, almost alone among mammals, were incapable of detecting sweet tastes. One said, without qualification, 'The cat shows no response to sweet tastes.' Another declared, 'Sweet tastes cannot be discerned by the cat.' This traditional wisdom must now be discarded. New tests have proved conclusively that cats *can* appreciate the presence of sweet tastes. If milk is diluted to one quarter of its normal strength, and hungry cats are then offered a choice between this weak milk laced with sucrose against the same milk without any sweetening, they always prefer the sweetened dishes.

If this is the case, why has it been denied in the past? The answer is that in most tests cats ignore the sweetness factor when making choices. It is of such minor significance to them that they 'override' it. If, for example, they are tested with full-strength or even half-strength milk, they show no preference for the more or the less sweetened examples. Their reaction to the milk itself is too strong. Only when the milk factor is greatly reduced by dilution does the sweetness factor begin to show. So although cats do enjoy this taste, they do so at a very mild level indeed.

Their strongest reaction is to sour tastes; next comes bitter, then salt and finally sweet. As food touches the tongue it comes into contact with the sensory papillae there. In the middle of the tongue these papillae are strong, rough and backward-pointing. In this area, there is a specialization of the tongue's surface that has nothing to do with taste. Indeed, there are no taste buds in this central region. It is a zone concerned entirely with rasping meat from bones or with cleaning fur. The taste buds are confined to the tip, the sides and the back of the tongue only. Sour tastes can be detected in all these areas, but bitter is confined to the back part and salt to the front.

However, the most powerful response of all to the food is to its smell, or fragrance. This is the really important information cats are receiving when they approach a meal. It is why many will sniff it and then walk away without even attempting to taste it. Like a wine connoisseur who only has to sniff the vintage to know how good it is, a cat can learn all it wants to know without actually trying the food.

If the animal does take a mouthful, then the tongue also has a sensitive reaction to the food's temperature. The wild ancestors of our domestic cats liked to eat freshly killed prey – they were not scavengers. And the tame descendants have kept the same views on this matter. The ideal, preferred temperature for feline food is 86°F, which happens to be the same temperature as the cat's tongue. Food taken straight from the refrigerator is anathema to the cat – unless it is very hungry, in which case it will eat almost anything. Sadly, for most cats today heated food is something of a luxury and, like many humans, they have learned to live with the 'fast food' mentality of modern times.

Why do cats sometimes reject their food?

Every cat owner knows the moment when a pet approaches a new dish of food, sniffs it, and then stalks off without taking a bite. It does not happen often, but when it does it is puzzling. Why should the cat suddenly refuse food that it normally eats with great enthusiasm?

Could the cat be ill? Yes, but this is not the explanation in many cases, where the animal is perfectly healthy in all other respects.

Could the food be bad? Yes, but again this cannot be the whole story because sometimes the food in question is identical to the previous meal, which was consumed greedily. Some owners have observed that two helpings of exactly the same cat food, given at different times on the same day, are treated differently – the first being eaten, the second rejected.

If the cat is well and the food is good, then what we are dealing with is a problem in feline behaviour, and any of a number of factors may be operating.

One explanation is that cats prefer to eat small meals on frequent occasions, rather than gorge on large, infrequent meals. Considering the size of their natural prey – small mice and birds – this is not so surprising. Unfortunately for domestic cats, their human owners rarely have the time to offer them mouse-sized meals, preferring to spoon out big dishfuls of cat food at feeding times. If you compare the amount of meat on an ordinary mouse with the amount of meat you place in your pet's food dish, you will find that the average cat-meal is about the equivalent of five mice. Although this is convenient for busy human owners, it is too much for the cat to eat at once, unless it is starving – which is rarely the case with well-loved family pets. Usually the cat eats a mouse-worth of food and then strolls off to digest it, returning later for another rodent-sized portion, and so on, until all the food is gone.

On this basis, if one meal is eaten and the dish emptied, and then the next meal is refused, it may simply be that the cat is not yet ready for its next 'kill'. Cats are normally extremely efficient at regulating the amount of food they take in. Obese cats are far less common than obese dogs (or obese people). So if they are being slightly overfed they may occasionally rebel and leave a new dish of food untouched.

Once again, this cannot be the complete story, because some owners have observed that their pets' daily intake is not always the same. On certain days the cats suddenly decide to eat much less than usual. Why should this be? One explanation is that sex is about to rear its head. If you have a female cat and it is coming into heat, it may temporarily go off its food, for example. Alternatively, if the weather suddenly becomes hotter or more humid, or both, cats may instantly cut down on their food intake.

Another possibility is that, unknown to you, your cat may be obtaining food elsewhere. A friendly neighbour may be giving titbits when visited by your roaming pet, which then returns home with its appetite ruined. Or there may have been a sudden and unexpected explosion in the local mouse population, leading to a spate of hunting and killing by a pet cat that normally feeds only at home. Again, its appetite will be dramatically reduced without any warning, leaving a perplexed owner to scrape a pile of stale cat food into the dustbin.

Another, less likely, possibility is that your cat hates the place in which it is given its food dish. Cats do not like to eat in a spot where there is bright light, a great deal of noise, or a lot of busy movement. They prefer to devour their 'prey' in a quiet, dim, private corner, away from the hustle and bustle of the home. Given an unsuitable feeding place, they may become erratic in their response to food. If they are unusually anxious or irritated, they may find the noise just too much to deal with and stalk off in a feline sulk rather than squat down for a good meal. Whether they eat or not in such cases will depend not so much on the food itself as on their varying mood.

Finally, even if all the factors mentioned so far are *not* influencing the cat, it may still turn its nose up at a particular dish of food. In such instances there is an inborn 'food variety mechanism' operating. This was originally discovered in birds, where seed-eaters were seen to switch from one type of seed to another from time to time, regardless of the nutritional similarity of the seeds. Given just one type of seed, they would always eat it and be perfectly healthy. They did not suffer from an unchanging diet. But if they were then given a choice of several kinds of seeds they would show sudden switches in preference, even though the seeds were chemically much the same. In the wild the importance of this mechanism is that it prevents a bird from becoming totally 'hooked' on one kind of foodstuff, so that if that type of seed suddenly disappears it is not left stranded. With wild cats it ensures that the animals do not become totally dependent on one kind of prey. With domestic cats it means that, every so often, the old 'standby' diet suddenly becomes unattractive and a brief change is required.

From the point of view of certain owners, these diet-shifts are nothing but a nuisance, but if ever, for some dramatic reason, the cat found itself without its usual owner, they would stand it in good stead. For they would enable the cat to switch more easily to whatever new food regimen was forced upon it by altered circumstances.

There is a seeming contradiction in cat feeding that must be mentioned here. If one cat is always given a completely monotonous (but nutritionally complete) diet, day in and day out, always the same brand of canned cat food, it may eventually refuse to touch any other, new kind of food, no matter how tasty. If another cat is given a much more interesting and varied diet, with a different kind of canned food each day and many other titbits, then, paradoxically, it may refuse one of its old, favourite foods from time to time. At first glance this does not make sense. The explanation is that, given a totally unvarying diet over a long period of time, especially if it is from kittenhood right through into adult life, a cat's 'food variety mechanism' gets worn down and is finally switched off altogether. It develops what is called 'neophobia', or 'fear of the new'. Novel tastes and smells become threatening. New foods are rejected. Such cats can become a real problem if their rigid daily routine is upset – say by the death of an elderly owner. The other type of cat – one that is given a more excitingly varied diet – has its 'food variety mechanism' fully activated throughout its life and therefore becomes more demanding and fussy with food. In other words, given no food choice a cat gives up asking for variety; given variety it demands more.

Can a cat survive on a vegetarian diet?

No, it is quite impossible for a cat – any cat – to survive on a vegetarian or vegan diet. Given a meatless diet it will rapidly become ill and will then die a painful death. A cat is a carnivore and if it is to be kept as a pet it must be given a carnivorous diet.

Nutritional experts have been criticized in the past for expressing themselves so strongly on this point, but they are unrepentant. If adult human beings wish to impose upon themselves an inappropriate and inefficient diet, that is their own business. But if they impose such a diet on their pet cats, they should be prosecuted for cruelty to animals.

This statement will anger many well-meaning vegetarians and vegans, but they must face the biological facts. Cats and humans both evolved as meat-eaters – as predators – and until our biochemists have been able to produce synthetic meats, with the magic mixture of essential amino-acids, we are both trapped by our evolutionary past. This does not reflect a lack of sympathy with the ethical basis of vegetarianism, far from it. Few people are happy about the idea of an animal having to die so that either they or their pet can can feed, and might well be prepared to eat a synthetic steak if one could be produced – and to serve it up to their cat. However, until that stage is reached, all we can do is to ensure that the animals whose flesh we devour are given the best possible lifestyle and are then despatched as quickly and painlessly as possible. If we cannot face the idea of feeding animal products to our pets then we should switch from keeping cats to keeping canaries.

For those who require more specific evidence to convince them, there are three key facts. Firstly, cats need an amino-acid called taurine to prevent them from going blind. Without it the retinas of their eyes would rapidly deteriorate. Some animals can manufacture taurine from other sources, but the cat cannot do so. It can only obtain it by eating animal proteins. So without a meat diet a cat would soon lose its sight.

Secondly, cats must have animal fats in their diets because they are incapable of manufacturing essential fatty acids without them. Some other animals can manage to convert vegetable oils into these fatty acids, but cats lack this ability. Without animal fats to eat, cats would, among other serious problems, find it difficult to achieve reproduction, blood-clotting and new cell production.

Thirdly, unlike many other animals, cats are unable to obtain vitamin A from plant sources (such as carrots) and must rely instead for this crucial substance on animal foods such as liver, kidney or fish oils.

These facts alone underline the folly of recent attempts to convert cat owners to vegetarian regimes for their pets.

Why do cats drink dirty water?

A number of owners have noticed, to their dismay, that their feline pets seem to have a passion for drinking from puddles and pools of water in the garden. They do this despite the fact that on the kitchen floor there is an immaculately clean dish of pure tap water, and probably milk as well, awaiting them. For some reason they ignore these hygienic delights and go padding off to some stagnant puddle to lap up the filthy water there. Why do they do it?

All the best books on cat care insist that cats should always have access to fresh, clean water and that the water should be changed regularly. They also tell their readers that the dish itself must be cleaned repeatedly to avoid any infection or contamination. But they overlook two problems. Fresh tap water is usually heavily treated with chemicals and often chlorinated strongly enough actually to have a chemical smell. The cat's sensitive nose cannot stand this. Worse still, the dish has probably been cleaned out with some form of modern detergent. This is bad enough on food dishes, but with those there is at least the powerful odour of the fish or meat to smother the distasteful detergent smell. With drinking dishes, however, the smell of the detergent is simply added to the already unpleasant odour of the treated water, with the result that the cat will only drink from its official dish if it has no alternative. The stale water in the puddles and pools outside is much more attractive. It may be full of microbes and rotting vegetation, but these are natural and organic and only give it an attractive flavour.

Veterinary authors issue dire warnings about the risk of 'transmission of disease' resulting from permitting your cat to drink from ponds and puddles, begging the question of how so many wild animals manage to remain perfectly healthy. In reality, the risk is really rather slight, though even this can be eliminated by taking more trouble to rinse off all the detergent from the cat's dishes. Because cats are many times more sensitive to detergent contamination than we humans are, much more rinsing than usual is required. Also, the fresh tap water should be allowed to stand for some time before being offered to the cat, to allow the chemicals to dissipate. Then, perhaps, the finicky felines will deign to dip their tongues into the clean water you offer them.

Also on the subject of drinking, a cat should never be given milk alone in place of water. If milk is offered, it should be given alongside water, so that the animal has a choice. Many adult cats, contrary to popular opinion, actively dislike milk, and it is not necessary for them. For some, especially Siamese cats, it can cause gastric upsets, and if milk is given without alternative water, such cats are liable to suffer from diarrhoea.

What is poisonous to cats?

Apart from the obvious poisons, cats are susceptible to a number of substances they may encounter in their daily lives. Nearly all of these are modern chemicals we have thoughtlessly introduced into our cats' environment to help us in various ways. In our urge to cleanse and control our world we have, for cats, often unwittingly polluted it.

The worst offenders are the disinfectants and pesticides. We may need them, but we should also spare a thought for our cats when using them, or sooner or later our pet animals will suffer. This applies both inside the house and outside, in the garden and on farmlands. Wild animals are not alone in suffering from some of the technological advances of modern agriculture.

One of the greatest hazards for today's free-roaming feline is the use of rodent poisons. Many people have employed a variety of mouse and rat poisons without stopping to think that the initial effect of such toxins is to slow down the victim which then becomes an easy target for a hunting cat. In other words, the dying mouse is the one most likely to end up in a cat's stomach. Inside the mouse's body, of course, the poison that made it so easy to catch is still present and can cause serious damage to the unfortunate cat. After eating its mouse, the cat may start to vomit, froth at the mouth and stagger about in a confused condition. Its heart beat may speed up or become weak, its breathing heavy and laboured, and it may eventually go into convulsions or start bleeding. If the rodent it ate had been heavily dosed, the cat may die. This is a case of adding injury to insult – it was originally the cat's job to kill the mice and the rodent-poisoners who usurped its role are not content merely with making it obsolete, but are also assaulting it physically in this underhand, prey-concealed way.

Another danger comes from domestic powders and sprays that are employed excessively – on the lawn for example. Cats that lie down on grass soaked in weed-killer and then fastidiously lick their fur clean will ingest this type of poison with alarming ease. Indoors, the various chemical insect-killers, disinfectants and furniture-cleaning sprays may also contaminate the pet cat's fur as the animal lies on the floor or on some other surface. Again it is the cat's obsessive cleanliness, driving it to groom its fur with its tongue, that accidentally transports these 'helpful' chemicals into its system. Small traces may not hurt, but in households where modern hygiene has become a fanatical pursuit, the unhappy pet is at some risk.

One of the mistakes we make is to imagine that if something is harmless or beneficial to us it must automatically be so to cats. Certain pain-killers we use – even something as mild and commonplace as aspirin – can be damaging to the cat. If we humanize the cat to the extent where we start to give it human medication, we may be hurting it when we are trying to be kind. If a cat is sick we should always take veterinary advice.

Sometimes, at Christmas or a special party, someone decides it would be fun to give the household cat a 'treat' by lacing its milk with alcohol. Most cats will refuse to join in such celebrations, but those that do will quickly suffer for it. Our human digestive system has a struggle with such things as alcohol, but we are resilient and are rather good at detoxifying the many dubious substances that we inflict on our long-suffering internal organs. The digestive system of the cat is less successful in this respect and often fails to break down the dangerous elements in ingested substances to make them harmless. Just as they cannot cope with the nutritional inferiority of a vegetarian diet, so they find it hard to deal with even a moderate dose of alcohol and may start to vomit, collapse or even go into a coma. Sobriety suits the domestic cat.

Finally there is one ancient, natural poison that can plague an inexperienced feline. The young adult cat, venturing into the garden in the spring, is excited to discover a prey that is mysteriously easy to catch. The common toad hops clumsily along and is incapable of making a sudden dart for cover. The cat is transfixed by this appealing sight and pounces, sinking its teeth into the prey. A moment later, the cat is in acute distress. Its mouth is turning red and swelling up. It starts to retch and dribble. Another over-eager cat has discovered one of the basic truths of garden life: never try to kill a toad.

The humble toad, so slow and cumbersome that it looks like

easy pickings for any predatory mammal or bird, has survived for millions of years because it has managed to evolve a particularly virulent poison called *bufotalin*. This is contained in the large warts that cover the upper surface of the toad's skin. These warts are no danger to a human being because they only exude their poisonous fluid if the toad is being injured. If the animal is bitten hard by a cat, or by some other unwary predator, the warts ooze poison into the attacker's clasping jaws. The toad is quickly dropped and escapes – with no more than a few minor punctures if it is lucky. It is significant that the two largest warts are positioned on either side of the top of its thick neck – precisely the place where a cat likes to sink its fangs when making its killing bite. In fact, these two warts are so enlarged that they look like long swellings behind the animal's eyes, covered with pores through which the poison seeps. Bearing in mind that it only takes 20 milligrams of this poison to kill a dog, and that a cat is equally susceptible, it is clear that here we have a serious threat to an incautious feline. Perhaps this is the origin of the old expression 'curiosity killed the cat'? Fortunately most pet cats, if they do encounter a toad, quickly learn that it tastes bad and, after one preliminary nip, rapidly drop the squirming amphibian and never make the same mistake again. Only the really savage hunter, who sinks his fangs deep into the toad's neck at its very first contact, will suffer a major trauma or death.

A final word of warning. Anti-freeze tastes sweet and cats like the flavour of it. If a pet cat has access to the garage where someone has been filling the car's radiator system with anti-freeze at the onset of winter, the animal may see a small puddle on the garage floor, where some has spilled. If it laps up a little of this, the ethylene glycol may cause irreversible kidney damage. The cat may even fall into a deep coma. The problem is that the pool of liquid usually forms underneath the car, where it is both inconspicuous and difficult to get at, so the chances are that it will not be mopped up. Unfortunately the space beneath a car is much more accessible to a cat's small body and the damage is easily done. Many times when a much-loved pet falls sick we think first of diseases and infections, when the real cause is simple chemical poisoning of a completely accidental nature.

Why do cats react so strongly to catnip?

In a word, it is because they are junkies. The catnip plant, a member of the mint family, contains an oil called *hepetalactone*, an unsaturated lactone which does for some cats what marijuana does for some people. When cats find this plant in a garden they take off on a ten-minute 'trip' during which they appear to enter a state of ecstasy. This is a somewhat anthropomorphic interpretation because we have no idea what is really happening inside the cat's brain, but anyone who has seen a strong catnip reaction will know just how trancelike and drugged the animal seems to become. All species of cats react in this way, even lions, but not every individual cat does so. There are some non-trippers and the difference is known to be genetic. With cats, you are either born a junkie or you are not. Conditioning has nothing to do with it. Under-age cats, incidentally, never trip. For the first two months of life all kittens avoid catnip, and the positive reaction to it does not appear until they are three months old. Then they split into two groups – those that no longer actively avoid catnip, but simply ignore it and treat it like any other plant in the garden, and those that go wild as soon as they contact it. The split is roughly 50/50, with slightly more in the positive group.

The positive reaction takes the following form: the cat approaches the catnip plant and sniffs it; then, with growing frenzy, it starts to lick it, bite it, chew it, rub against it repeatedly with its cheek and its chin, head-shake, rub it with its body, purr loudly, growl, miaow, roll over and even leap in the air. Washing and clawing are also sometimes observed. Even the most reserved of cats seems to be totally disinhibited by the catnip chemical.

Because the rolling behaviour seen during the trancelike state is similar to the body actions of female cats in oestrus, it has been suggested that catnip is a kind of feline aphrodisiac. This is not particularly convincing, because the 50 per cent of cats that show the full reaction include both males and females, and both entire animals and those which have been castrated or spayed. So it does not seem to be a 'sex trip', but rather a drug trip which produces similar states of ecstasy to those experienced during the peak of sexual activity.

Cat junkies are lucky. Unlike so many human drugs, catnip does no lasting damage, and after the ten-minute experience is over the cat is back to normal with no ill-effects.

Catnip (*Nepeta cataria*) is not the only plant to produce these strange reactions in cats. Valerian (*Valeriana officinalis*) is another one, and there are several more that have strong cat-appeal. The strangest discovery, which seems to make no sense at all, is that if catnip or valerian are administered to cats internally they act as tranquillizers. How they can be 'uppers' externally and 'downers' internally remains a mystery.

Why does a cat wag its tail when it is hunting a bird on a lawn?

The scene is familiar to most cat-owners. Through the window they see their cat stalking a bird by creeping stealthily towards it, head down and body low on the ground. This cautious crouching attempt to be as inconspicuous as possible is suddenly and dramatically ruined by the animal's tail, which starts swishing uncontrollably back and forth through the air. Such movement acts like a flag being waved at the bird to warn it of approaching danger. The intended victim takes off immediately and flies to safety, leaving a frustrated feline hunter staring up into the sky.

Cat-owners witnessing this scene are puzzled by their cat's inefficiency. Why does the cat's tail betray the rest of the body in this self-defeating way? Surely the wild ancestors of the domestic cat could not have survived such a serious flaw in their hunting technique? We know that conspicuous tail-wagging in cats is a social signal indicating acute conflict. It is useful when employed between one cat and another and is then an important part of feline body language. But when it is transferred into a hunting context, where the only eyes that will spot the signal are those of the intended prey, it wrecks the whole enterprise. So why has it not been suppressed in such cases?

To find the answer we have to look at the normal hunting sequence of the cat. This does not take place on an open lawn and is less well known to cat-owners than it might be because it involves a great deal of waiting and hiding. If owners do happen on a hunt in progress they will automatically disrupt it, so that there is nothing more to observe. The disturbed prey escapes and the cat gives up. So for a casual observer the whole sequence is not easy to study. It requires some systematic and secretive catwatching. When this is undertaken, the following points emerge:

First, the cat makes a great deal of use of cover. It spends much time lying half-hidden in undergrowth, often with only its eyes and part of its face visible. The tail is usually completely hidden from view. Second, it never attempts to pounce on a prey until it is very close to it. It is not a prey-chaser. It may make a few stalking-runs, rushing forward in its flattened posture, but it then halts and waits again before pouncing. Third, its normal prey is not birds but rodents. A careful study of feral cats in the United States revealed that birds accounted for only 4 per cent of the diet. The excellent eyesight of birds and their ability to fly straight up in the air to escape make them unsuitable targets for domestic cats.

Together these points explain the dilemma of the suburban cat hunting a bird on a lawn. To start with, the open, manicured lawn robs the cat of all its natural cover, exposing its whole body to view. This is doubly damaging to its chances. It makes it almost impossible for the cat to creep near enough for its typical, close-quarters pounce without being seen. This puts it into an acute conflict between wanting to stay immobile and crouched, on the one hand, and wanting to rush forward and attack, on the other. The conflict starts its tail wagging furiously and the same lack of cover that created the conflict then cruelly exposes the vigorous tail movements to the frightened gaze of the intended prey.

If the attempt to hunt a bird on an open lawn is so doomed to failure, why does the cat keep trying? The answer is that every cat has a powerful urge to go hunting at regular intervals, but this urge has been severely hampered by advances in human pest control. In town and cities and suburbs, the rodent population that used to infest houses and other dwellings has been decimated by modern techniques. Garden birds may be pests, but their appeal to human eyes has protected them from a similar slaughter. As a result, the rodent-hunting cat finds itself today in an unnaturally mouse-free, bird-rich environment. It cannot utilize its natural hunting skills under such conditions. It is this state of affairs that drives the cat on to crouch hopelessly but compulsively on open lawns, staring longingly at elusive birds. So when it waves its tail at its prey on these occasions, it is not the cat which is a stupid hunter, but we who have unwittingly forced a clever hunter to attempt an almost impossible task.

Why does a cat chatter its teeth when it sees a bird through the window?

Not every owner has observed this curious action, but it is so strange that it is a case of 'once seen never forgotten'. The cat, sitting on a window-sill, spots a small bird conspicuously hopping about outside and stares at it intently. As it does so it begins juddering its teeth in a jaw movement which has variously been described as a 'tooth-rattling stutter', a 'tetanic reaction' and 'the frustrated chatter of the cat's jaws in the mechanical staccato fashion'. What does it mean?

This is what is known as a 'vacuum activity'. The cat is performing its highly specialized killing-bite, as if it already had the unfortunate bird clamped between its jaws. Careful observation of the way in which cats kill their prey has revealed that there is a peculiar jaw movement employed to bring about an almost instantaneous death. This is important to a feline predator because even the most timid of prey may lash out when actually seized, and it is vital for the cat to reduce as much as possible any risk of injury to itself from the sharp beak of a bird or the powerful teeth of a rodent. So there is no time to lose. After the initial pounce, in which the prey is pinioned by the strong claws of the killer's front feet, the cat quickly crunches down with its long canine teeth, aiming at the nape of the neck. With a rapid juddering movement of the jaws it inserts these canines into the neck, slipping them down between vertebrae to sever the spinal cord. This killing-bite immediately incapacitates the prey and it is an enactment of this special movement that the frustrated, window-gazing cat is performing, unable to control itself at the tantalizing view of the juicy little bird outside.

Incidentally, this killing-bite is guided by the indentation of the body outline of the prey – the indentation which occurs where the body joins the head in both small birds and small rodents. Some prey have developed a defensive tactic in which they hunch up their bodies to conceal this indentation and in this way make the cat miss its aim. If the trick works, the cat may bite its victim in part of the body which does not cause death, and on rare occasions the wounded prey may then be able to scrabble to safety if the cat relaxes for a moment, imagining that it has already dealt its lethal blow.

Why does a cat sway its head from side to side when staring at its prey?

When a cat is preparing to pounce on its prey it sometimes sways its head rhythmically from side to side. This is a device employed by many predators blessed with binocular vision. The head-sway is a method of checking the precise distance at which the prey is located. If you sway your own head from side to side you will see how, the closer an object is, the more it is displaced by the lateral movements. The cat does this to refine its judgment, because when the rapid pounce forward is made it must be inch-accurate or it will fail.

Why does a cat sometimes play with its prey before killing it?

Horrified cat-owners have often experienced the shock of finding their pet cats apparently torturing a mouse or small bird. The hunter, instead of delivering the killing-bite of which it is perfectly capable, indulges in a cruel game of either hit-and-chase or trap-and-release, as a result of which the petrified victim may actually die of shock before the final *coup de grâce* can be delivered. Why does the cat do it, when it is such an efficient killing machine?

To start with, this is not the behaviour of a wild cat. It is the act of a well-fed pet which has been starved of hunting activity as a consequence of the 'hygienic' environment in which it now lives – neat suburbs or well-kept villages where the rodent infestation has long since been dealt with by poison and human pest control agencies. For such a cat, the occasional catching of a little field mouse, or a small bird, is a great event. It cannot bear to end the chase, prolonging it as much as possible until the prey dies. The hunting drive is independent of the hunger drive – as any cat-owner knows whose cat has chased off after a bird on the lawn immediately after filling its belly with canned cat food. Just as hunger increases without food, so the urge to hunt increases without access to prey. The pet cat over-reacts, and the prey suffers a slow death as a result.

On this basis, one would not expect feral cats which are living rough or farm cats employed as 'professional pest-controllers' to indulge in play with their half-dead prey. In most cases it is indeed absent, but some researchers have found that female farm cats do occasionally indulge in it. There is a special explanation in their case. As females, they will have to bring live prey back to the nest to demonstrate killing to their kittens at a certain stage in the litter's development. This maternal teaching process will account for an eagerness on the part of females to play with prey even though they are not starved of the hunting process.

There is one other explanation for this seemingly cruel behaviour. When cats attack rats they are quite nervous of their prey's ability to defend themselves. A large rat can give a nasty bite to a cat and has to be subdued before any attempt is made to perform the killing-bite. This is done by the cat swinging a lightning blow with its claws extended. In quick succession it may beat a rat this way and that until it is dazed and dizzy. Only then does the cat risk going in close with its face for the killing-bite. Sometimes a hunting cat will treat a small mouse as if it were a threatening rat, and start beating it with its paws instead of biting it. In the case of a mouse this quickly leads to a disproportionately savage pounding, with the diminutive rodent being flung to and fro. Feline behaviour of this type may appear like playing with the prey, but it is distinct from the trap-and-release play and should not be confused with it. In trap-and-release play the cat is inhibiting its bite each time. It is genuinely holding back to prolong the hunt. In hit-and-chase attacks on mice the cat is simply over-reacting to the possible danger from the prey's teeth. It may look like cruel play, but in reality it is the behaviour of a cat that is not too sure of itself. Even after the prey is nearly or completely dead, such a cat may continue to bat the victim's body around, watching it intently to see if there is any sign of retaliation. Only after a long bout of this treatment will the cat decide that it is safe to deliver the killing-bite and eat the prey. An experienced, full-time hunter would not react in this way, but a pampered pet cat, being a little rusty on the techniques of a quick kill, may well prefer this safer option.

How does a cat prepare its food?

Immediately after the kill, a cat goes through the strange little routine of 'taking a walk'. Unless it is starving, it paces up and down for a while, as if feeling the need to release the tension of the hunt-and-kill sequence. Only then does it settle down to eating the prey. This pause may be important for the cat's digestion, giving its system a chance to calm down after the adrenalin-excitement of the moments that have just passed. During this pause a prey that has been feigning death may try to escape and, on very rare occasions, succeeds in doing so before the cat can return to the hunting mood again.

When the cat finally approaches its prey to eat it, there is the problem of how to prepare it for easy swallowing. Small rodents cause no difficulties. They are simply eaten head first and the skins, if swallowed, are regurgitated later. Some cats separate out the gall-bladder and intestines and avoid eating them, but others are too hungry to care and gobble down the entire animal without any fuss.

Birds are another matter because of their feathers, but even here the smaller species are eaten in their entirety, with the exception of tail and wing feathers. Birds the size of thrushes and blackbirds are plucked a little before eating, but then the cat impatiently starts its meal. After a while it breaks off to remove a few more feathers, before eating further. It repeats this a number of times as the feeding proceeds. Bigger birds, however, demand more systematic plucking, and if a cat is successful at killing a pigeon or something larger, it must strip away the feathers before it begins to eat.

To pluck a pigeon, a cat must first hold down the body of the bird with its front feet, seize a clump of feathers between its teeth, pull its jaw-clamped head upwards with some force, and then finally open its mouth and shake its head vigorously from side to side to remove any clinging plumage. As it shakes its head it spits hard and makes special licking-out movements with its tongue, trying to clear its mouth of stubbornly attached feathers. It may pause from time to time to lick its flank fur. This last action puts grooming into reverse. Normally the tongue cleans the fur, but here the fur cleans the tongue. Any last remnants are removed and then the next plucking action can take place.

The urge to pluck feathers from a large bird appears to be inborn. I once presented a dead pigeon to a wild cat living in a zoo cage where it had always been given chunks of meat as its regular diet. The cat became so excited at seeing a fully feathered bird that it started an ecstatic plucking session that went on and on until the whole body of the bird was completely naked. Instead of settling down to eat it, the cat then turned its attention to the grass on which it was sitting and began plucking that. Time and again it tugged out tufts of grass from the turf and shook them away with the characteristic bird-plucking movements until, eventually, having exhausted its long-frustrated urge to prepare its food, the cat finally bit into the flesh of the pigeon and began its meal. Clearly, plucking has its own motivation and can be frustrated by captivity, just like other, more obvious drives.

The strangest feature of feather-plucking is that Old World Cats perform it differently from New World Cats. All species from the first area perform a zigzag tugging movement leading to the full shake of the head, while those from the Americas tug the feathers out in a long vertical movement, straight up, and only then perform the sideways shake. It appears that, despite superficial similarities between the small cats from the two sides of the Atlantic, they are in reality two quite distinct groups.

How efficient is the cat as a pest-killer?

Before the cat became elevated to the level of a companion and pet for friendly humans, the contract between man and cat was based on the animal's ability to destroy pests. From the time mankind first started to keep grain in storage, the cat had a role to play and carried out its side of the bargain with great success.

Not so long ago it was thought that the best way to get farm cats to kill rats and other rodent pests was to keep the feline hunters as hungry as possible. This seemed obvious enough, but it was wrong. Hungry farm cats spread out over a huge hunting territory in search of food and killed fewer of the pests inside the farm. Cats that were fed by the farmer stayed nearer home and their tally of farm pests was much higher. The fact that they had been fed already and were not particularly hungry made no difference to the number of prey they killed each day, because the urge to hunt is independent of the urge to eat. Cats hunt for the sake of hunting. Once farmers realized this they were able to keep their cats close by the farm and reduce the damage done to their stores by rodent pests. A small group of farm cats, well looked after, could prevent any increase in the rodent population, providing a major infestation had not been allowed to develop before their arrival.

According to one authority, the champion mouser on record was a male tabby living in a Lancashire factory where, over a very long lifespan of twenty-three years, he killed more than 22,000 mice. This is nearly three a day, which seems a reasonable daily diet for a domestic cat, allowing for some supplements from human friends, but it is far exceeded by the world's champion ratter. That honour goes to a female tabby which earned her keep at the late lamented White City Stadium. Over a period of only six years she caught no fewer than 12,480 rats, which works out at a daily average of five to six. This is a formidable achievement and it is easy to see why the ancient Egyptians went to the trouble of domesticating cats and why the act of killing one was punishable by death.

Why do cats present freshly caught prey to their human owners?

They do this because they consider their owners such hopeless hunters. Although usually they look upon humans as pseudo-parents, on these occasions they view them as their family – in other words, their kittens. If kittens do not know how to catch and eat mice and small birds, then the cat must demonstrate to them. This is why the cats that most commonly bring home prey and offer such gifts to their owners are neutered females. They are unable to perform this action for their own litters, so they redirect it towards human companions.

The humans honoured in this way frequently recoil in horror or anger, especially if the small rodent or bird is still half-alive and struggling. The cat is totally nonplussed by this extra-ordinary response. If it is scolded for its generous act, it once again finds its human friends incomprehensible. The correct reaction would be to praise the cat for its maternal generosity, take the prey from it with many compliments and strokings and then quietly dispose of it.

Under natural conditions a cat which has a litter of kittens introduces them to prey animals little by little. When they are about seven weeks old, instead of killing and eating her prey where she catches it, she kills it and then brings it back to where the kittens are kept. There she proceeds to eat it while they watch. The next phase involves bringing the dead prey back and playing with it before consuming it, so that the kittens can see her beating it with her claws and grabbing it. The third phase involves leaving the prey to be eaten by the kittens themselves. But she is still not prepared to risk bringing a live or even a half-dead prey to the kittens, because it could easily bite them or attack them if they are unwary. Only when they are a little older will she do this, and then she herself will make the kill in front of the kittens. They watch and learn. Eventually they will accompany her on the hunt and try killing for themselves.

Why do cats eat grass?

Most cat-owners have observed the way in which, once in a while, their pet goes up to a long grass stem in the garden and starts to chew and bite at it. Cats living in apartments where there are no gardens in which to roam have been known to cause considerable damage to house plants in desperate attempts to find a substitute for grasses. In rare cases such cats have even harmed themselves by biting into plants that are poisonous.

Many cat experts have puzzled over this behaviour and some have admitted frankly that they have no answer. Others have offered a variety of explanations. For many years the favourite reply was that the cats use grass as a laxative to help them pass troublesome hairballs lodged in their intestines. A related suggestion claimed that they were eating grass to make themselves vomit up the hairballs. This was based on the observation that cats do sometimes vomit after eating grass, but it overlooked the possibility that whatever made the cats feel sick also made them want to eat grass, rather than that the grass-eating actually caused the vomiting.

A less popular explanation was that the grass aided the cats in the case of throat inflammation, or irritation of the stomach. Some authorities simply dismissed the activity as a way of adding roughage to the diet.

None of these explanations makes much sense. The amount of grass actually eaten is very small. Watching the cats chewing at long grasses, one gets the impression that they are merely taking in a little juice from the leaves and stems, rather than adding any appreciable solid bulk to their diet.

The most recent opinion – and the most likely explanation – is that cats chew grass to obtain minute quantities of a chemical substance that they cannot obtain from a meat diet and which is essential to their health. The substance in question is a vitamin called folic acid, and it is vital to cats because it plays an important role in the production of haemoglobin. If a cat is deficient in folic acid its growth will suffer and it may become seriously anaemic. Cat-owners whose animals have no access to grasses of any kind sometimes solve the problem by planting grass seeds in a tray and growing a patch of long grass in their apartments for their pets to chew on.

Incidentally, it is worth pointing out that although cats may need this plant supplement to their meat diet, they are first and foremost carnivores and must be treated as such. Recent attempts by well-meaning vegetarians to convert their cats to a meat-free diet are both misguided and cruel. Cats rapidly become seriously ill on a vegetarian diet and cannot survive it for long. The recent publication of vegetarian diets recommended as suitable for cats is a clear case of animal abuse and should be dealt with as such.

Why does a cat bury its faeces?

This action is always referred to as an indication of the fastidious tidiness of the cat. Owners of messy dogs are often regaled with this fact by cat-owners insisting on the superiority of felines over canines. This favoured interpretation of faeces-burying as a sign of cat hygiene does not, however, stand up to close investigation.

The truth is that cats bury their faeces as a way of damping down their odour display. Faeces-burying is the act of a subordinate cat, fearful of its social standing. Proof of this was found when the social lives of feral cats were examined closely. It was discovered that dominant tom-cats, far from burying their faeces, actually placed them on little 'advertising' hillocks, or any other raised points in the environment where the odour could be wafted abroad to maximum effect. It was only the weaker, more subdued cats which hid their faeces. The fact that our pet cats always seem to carry out such a careful burying routine is a measure of the extent to which they see themselves dominated by us (and also perhaps by the other cats in the neighbourhood). This is not really so surprising. We are physically stronger than they are and we completely dominate that all-important element in feline life – the food supply. Our dominance is in existence from the time of kittenhood onwards, and never in serious doubt. Even big cats, such as lions, can be kept in this subordinate role for a lifetime, by their friendly owners, so it is hardly surprising that the small domestic cat is permanently in awe of us and therefore always makes sure to bury its faeces.

Burying the faeces does not, of course, completely switch off the odour signal, but it does reduce it drastically. In this way the cat can continue to announce its presence through its scents, but not to the extent that it transmits a serious threat.

Why do some cats foul the house?

Some owners are distressed to discover that, after months or years of 'clean' behaviour, their pet cats start to leave messes around the house. In the past such animals have been models of cleanliness, always asking to go outside before defecating, or confining themselves to a modern litter tray. Now suddenly, for no apparent reason, they become careless and leave their faeces on the carpet. If there is no obvious explanation for this loss of restraint, what is the hidden one?

The first possibility is that the cat in question is ill or senile and its whole behaviour routine is starting to collapse. In such cases a vet is needed urgently. But many cats who foul the house are clearly perfectly healthy, at least physically. So what is their mental state, that causes them to behave in this very unfeline-like way?

One factor that is important could be classified as 'territorial disruption'. Some people find that the trouble begins when they have the builders in and the house undergoes some kind of dramatic redesigning. Cats hate this and, if a new wing has been added to the home for example, will defecate on the floor of the new area, almost as a comment on how they feel about the upheaval. One owner found that his cat did this only on the floor of a new extension he had built on to the side of his house. The rest of the dwelling was treated with as much respect as before, but the new room was — as far as the cat was concerned — still part of the 'outside' and was treated accordingly.

A second factor is 'social disruption'. If a new kitten is brought into the home the old cat may feel slighted by all the attention the newcomer is receiving. It will feel the need to express its dominance and will do this in the manner of a 'top cat'. This means leaving faeces in a prominent place, rather than covering them up in the litter tray. Most people still imagine that all cats cover their droppings, but this is not so. Only friendly or subordinate cats cover them up. Aggressively dominant, wild-living cats leave them openly in conspicuous places as 'odour threats'. In an undisturbed home, all domestic cats see themselves as subordinates of their human owners, so under normal circumstances all domestic cats use litter trays or bury their faeces in the garden. But if they suddenly have the urge to express their senior status in relation to a new pet in the home they may revert to the age-old manner of doing this, much to the distress of their well-meaning owners, who probably bought the new kitten as a special companion for the older cat.

A third possibility, and perhaps the most likely explanation, is that the cat's litter tray leaves something to be desired. Cats hate to bury their faeces in a place where they have recently done so. In the garden you will see them nosing around trying to find a new place to dig a hole. In the litter tray they do the same thing, but if it has been used several times without being properly cleaned out this becomes impossible and the cat will then prefer to defecate elsewhere, even if it has to go through the motions of covering its dung with imaginary earth after it has deposited it on a wooden floor or a carpet. The secret in such cases is to increase the rate of emptying and cleaning the litter tray.

A final factor has to do with the position of the litter tray. Cats hate to defecate where they eat and some people place the litter tray too near the animal's food dish. Or they place it in a busy part of the house, near the back door, where people are always walking past. Cats feel vulnerable when they are defecating and do not like to have anyone near them at that time. So if a litter tray is in too public a place, this too may drive them elsewhere.

If the trouble has started and does not look like stopping, the best solution is to provide an immaculately clean tray of fresh litter, place it in a quiet corner, and then shut the animal up in that particular room, releasing it only when it has used the tray in the approved manner. In this way it is possible to re-start the routine that was enjoyed previously, although it may take some days of patience to break the new habit.

What smells actively repel cats?

The chemical smells of modern tap water may stop a cat drinking from its bowl, but they are not strong enough to drive the animal back physically from the corner of the kitchen floor where the bowl is placed. Odours offensive enough to put a cat into full retreat are rare. This creates a problem for people who are looking for something that cats dislike, to use as a repellent. When a pet cat has started scratching the fabric of a valuable chair, for example, or has begun to make messes on an expensive carpet, it would be helpful to be able to sprinkle or smear some hated odour there to keep the animal away. But what can be used?

Searching back through the long history of feline deterrents, there appear to have been only three smelly substances that have achieved a measure of success. The first is the oil of the crushed leaves of a small aromatic bush called 'rue'. As long ago as the first century A.D., the Roman author Pliny, in his monumental Natural History, suggested that placing branches of this shrub around an object would keep cats away from it. This advice was still being offered 1,200 years later, in the Middle Ages, when an expert on herb gardens wrote, 'Behind the turf plot let there be a great diversity of medicinal and aromatic herbs, among which Rue should be mingled in many places for its beauty and greenness, and its bitterness will drive away poisonous animals from the garden.' Some modern gardeners report that handling the leaves of this plant can cause a blistering rash on sensitive skin, so it must obviously be treated with respect, but there is a good chance that oil of rue would prove to be a successful repellent for most misbehaving cats. For some reason this ancient piece of folk-wisdom seems to have been largely forgotten, but it might be worth reviving it in cases where ordinary measures have failed.

A second and perhaps easier suggestion is the use of onions. By rubbing a raw onion over the area to be protected, a cat can be deterred, and the smell, although unpleasant for humans at first, soon goes unnoticed. Most cats, however, continue to find it distasteful long after the human occupants of the rooms have forgotten about it.

The most effective deterrent, however, is that simple household substance, vinegar. Cats loathe it. The acid fragrance upsets their delicate nasal passages and they avoid anything smeared in it for long periods of time. Short of buying special, commercially prepared cat-repellent sprays from a local pet shop, it is the best weapon available.

It must be added, though, that cats are stubborn animals and will often consider such chemical warfare as a kind of challenge. Their first reaction will be to shift the location of their activities. If this fails, after a while they may manage to overcome their distaste for a particular substance and then it is necessary to change one's tactics. Ultimately, the best solution is to try and defeat the cat's damaging ways by using understanding and intelligence rather than foul-smelling chemicals. If the behaviour behind the unwanted feline activity can be analysed, it may be possible to find a psychological solution which, in the long run, will always be more successful.

Note:

Since completing this book I have been able to obtain a cutting of *rue* and test its effect on my own cat. Placed on the carpet, it attracted her and she sat sniffing it closely. But when I rubbed the leaves between my fingers and offered her my fingertips to sniff, her reaction was dramatic. She brought her nose up to my hand, then leapt backwards, opened her mouth and tried to vomit. After that she stalked off and refused to come near me. I washed my hands and tried to make my peace with her, but when I went to stroke her she hissed at me. Ten minutes later, she miaowed if I approached her and it took her several hours to forgive me and to stop treating me like a walking cat repellent.

Why does a tom-cat spray urine on the garden wall?

Tom-cats mark their territories by squirting a powerful jet of urine backwards on to vertical features of their environment. They aim at walls, bushes, tree-stumps, fence-posts or any landmark of a permanent kind. They are particularly attracted to places where they or other cats have sprayed in the past, adding their new odour to the traces of the old ones already clinging there.

The urine of tom-cats is notoriously strongly scented, so much so that even the inefficient human nostrils can detect it all too clearly. To the human nose it has a particularly unpleasant stench and many people are driven to having toms neutered in attempts to damp down this activity. Other cat odours are almost undetectable by humans. The glands on the head, for example, which are rubbed against objects to deposit another form of feline scent-mark, produce an odour that is of great significance to cats but goes completely unnoticed by the animals' human owners.

Some authorities have claimed that the squirted urine acts as a threat signal to rival cats. Hard evidence is lacking, however, and many hours of patient field-observation have never revealed any reactions to support this view. If the odour left on the landmarks was truly threatening to other cats, it should intimidate them when they sniff it. They should recoil in fear and panic and slink away. Their response is just the opposite. Instead of withdrawing, they are positively attracted to the scent-marks and sniff at them with great interest.

If they are not threatening, what do the territorial scent-marks signify? What signals do they carry? The answer is that they function rather as newspapers do for us. Each morning we read our papers and keep up to date with what is going on in the human world. Cats wander around their territories and, by sniffing at the scent-marks, can learn all the news about the comings and goings of the feline population. They can check how long it has been since their own last visit (by the degree of weakening of their own last scent-spray) and they can read the odour-signs of who else has passed by and sprayed, and how long ago. Each spray also carries with it considerable information about the emotional state and the individual identity of the sprayer. When a cat decides to have another spray itself, it is the feline equivalent of writing a letter to *The Times*, publishing a poem, and leaving a calling card, all rolled into one jet of urine.

It might be argued that the concept of scent-signalling is far-fetched and that urine-spraying by cats is simply their method of getting rid of waste products from the body and that it has no other significance whatsoever. If a cat has a full bladder it will spray; if it has an empty bladder it will not spray. The facts contradict this. Careful observation shows that cats perform regular spraying actions in a set routine, regardless of the state of their bladder. If it happens to be full, then each squirt is large. If it is nearly empty, then the urine is rationed out. The *number* of squirts and the territorial areas which are scent-marked remain the same, no matter how much or how little liquid the cat has drunk. Indeed, if the cat has completely run out of urine, it can be seen continuing its scent-marking routine, laboriously visiting each scent-post, turning its back on it, straining and quivering its tail, and then walking away. The act of spraying has its own separate motivation, which is a clear indication of its importance in feline social life.

Although it is not generally realized, females and neutered cats of both sexes do spray jets of urine, like tom-cats. The difference is that their actions are less frequent and their scent far less pungent, so that we barely notice it.

Why does a cat spend so much time grooming its fur?

The obvious answer is to keep itself clean, but there is much more to grooming than this. In addition to cleaning away dust and dirt or the remains of the last meal, the repeated licking of the fur helps to smooth it so that it acts as a more efficient insulating layer. A ruffled coat is a poor insulator, which can be a serious hazard for a cat in freezing weather.

Cold is not the only problem. Cats easily overheat in summer-time and fur-grooming increases then for a special reason. Cats do not have sweat glands all over their bodies as we do, so they cannot use sweating as a rapid method of cooling. Panting helps, but it is not enough. The solution is to lick repeatedly at the fur and deposit on it as much saliva as possible. The evaporation of this saliva then acts in the same way as the evaporation of sweat on our skin.

If cats have been in sunlight they increase their grooming even more. This is not, as might be imagined, simply because they are even hotter, but because the action of sunlight on their fur produces essential vitamin D. They acquire this crucial additive to their diet by the licking movements of their tongues over the sun-warmed fur.

Grooming also increases when cats become agitated. This is called displacement grooming and it is believed to act as an aid to relieving the strain of tense social encounters. When we are in a state of conflict we often 'scratch our heads'. A cat under similar conditions licks its fur.

Any cat-owner who has just been holding or cuddling their cat will be familiar with the animal's actions as soon as it has been released from human contact. It wanders off, sits down and then, nearly always, starts to groom itself. This is partly because it needs to smooth its ruffled fur, but there is also another reason. You have, by handling the cat, given it your scent and to some extent masked the cat's scent. The licking of the fur redresses the balance, weakening your scent and reinforcing the cat's own odour on its body surface. Our lives are dominated by visual signals, but in the cat's world odours and fragrances are much more important, and an overdose of human scent on its fur is disturbing and has to be rapidly corrected. In addition, the licking of the fur you have been handling means that the cat can actually enjoy 'tasting' you and reading the signals it gets from the scent of your sweat glands. *We* may not be able to smell the odour of our hands, but a cat can.

Finally, the vigorous tugging at the fur which is so typical of a cat's self-grooming actions plays a special role in stimulating the skin glands at the base of the individual hairs. The secretions of these glands are vital to keep the fur water-proofed, and the tugging of the cat's busy tongue steps up the waterproofing as a protection against the rain.

So grooming is much more than mere cleaning. When it licks its fur a cat is protecting itself, not only from dirt and disease, but also from exposure to cold, from overheating, from vitamin deficiency, from social tension, from foreign odours and from getting drenched to the skin. No wonder it devotes so much of its waking day to this piece of behaviour.

There is one danger inherent in this activity. Moulting cats and cats with very long fur quickly accumulate a large number of hairs inside their alimentary tracts and these form into hairballs which can cause obstructions. Usually hairballs are vomited up naturally without causing any trouble, but if they grow too large they may become a serious hazard. Cats of a nervous disposition, which do a great deal of displacement grooming, also suffer in this way. To solve their problem it is necessary to find out what is causing their agitation and deal with it. For the moulting and long-haired cats the only prevention is regular grooming by the cat's owner with brush and comb, to remove the excess fur.

Self-grooming begins when the kitten is about three weeks old, but it has its fur attended to by the mother from the moment of its birth. Being groomed by another cat is called allogrooming, in contrast with self-grooming which is known technically as autogrooming. Allogrooming is common not only between mother and kitten, but also between adult cats that have grown up together and have developed a close social bond. Its primary function is not mutual hygiene, but rather a cementing of the friendly relation that exists between the two animals. All the same, licking in a region that is hard for the cat itself to reach does have a special appeal, and cats are partial to attention behind the ears. This is why tickling and rubbing behind the ears is such a popular form of contact between cat-owners and their cats.

The autogrooming actions often follow a set sequence, when a cat is indulging in a complete 'wash-and-brush-up'. The typical routine goes as follows:

1 Lick the lips.
2 Lick the side of one paw until it is wet.
3 Rub the wet paw over the head, including ear, eye, cheek and chin.
4 Wet the other paw in the same way.
5 Rub the wet paw over that side of the head.
6 Lick front legs and shoulders.
7 Lick flanks.
8 Lick genitals.
9 Lick hind legs.
10 Lick tail from base to tip.

If at any stage during this process an obstruction is encountered – a tangled bit of fur, for example – the licking is momentarily abandoned in favour of a localized nibble with the teeth. Then, when all is clear, the grooming sequence is resumed. Foot and claw nibbling are particularly common, removing dirt and improving the condition of the claws. This complicated cleaning sequence differs from that seen in many other mammals. Rats and mice, for example, use the whole of their front paws for grooming their heads, whereas the cat uses only the side of the paw and part of the forearm. Also, rodents sit up on their back legs and groom with both front feet at the same time, while the feline technique is to employ each front leg alternately, resting its body on the one not in use. Human observers rarely comment on such differences, remarking simply that an animal is busy cleaning itself. In reality, closer observation reveals that each species follows a characteristic and complex sequence of actions.

How many kinds of hairs do cats have?

A wild cat has four kinds: down hairs, awn hairs, guard hairs and vibrissae. There may be as many as two hundred hairs per square millimetre, giving the cat an excellent fur coat that can protect it from even the most severe night air.

The *down hairs* are the ones closest to the skin and it is their primary task to keep the animal warm and to conserve its precious body heat. These are the shortest, thinnest and softest of the hairs. They have roughly the same diameter down their whole length, but instead of being straight they have many short undulations, making them appear crimped or crinkled when viewed under a magnifying lens. It is the soft and curly quality of this undercoat, or underfur, that gives it its excellent heat-retaining property.

The *awn hairs* form the middle-coat. They are intermediate between the soft underfur and the guard hairs of the topcoat. Their task is partly insulatory and partly protective. They are bristly with a slight swelling towards the tip, before the final tapering-off. Some authorities subdivide them into three types – the down-awn hairs, the awn hairs and the guard-awn hairs – but these subtle distinctions are of little value.

The *guard hairs* form the protective topcoat. They are the longest and thickest of the ordinary body hairs and serve to protect the underfur from the outside elements, keeping it dry and snug. These hairs are straight and evenly tapered along their length.

The *vibrissae* are the greatly enlarged and toughened hairs employed as sensitive organs of touch. These specialized tactile hairs form the whiskers of the upper lips, and are also found on the cheeks and the chin, over the eyes and on the wrists of the forelegs. Compared with the other types of hair there are very few of them, but they play a vital role when the cat is exploring in poor light, or is hunting.

Of the three types of general body fur on the wild cat, the down hairs are the most numerous. For every thousand down hairs there will only be about three hundred awn hairs and twenty guard hairs. But these ratios vary enormously with the different breeds of pedigree cats. This is because these felines have been carefully selected for their special kinds of coats. Some are fine and thin, others short and coarse, or long and fluffy. The differences are due to exaggerations and reductions of the different types of hair.

Pedigree long-haired cats, for example, have excessively lengthy guard hairs, measuring up to 5 inches, and greatly elongated down hairs, but no awn hairs. Some short-haired breeds have guard hairs that are less than 2 inches in length, sparse awn hairs and no down hairs. Wirehair cats have all three types of body hair, but they are all short and curly. The strange Cornish Rex cats have no guard hairs and only very short, curly awn and down hairs. The Devon Rex has all three types of body hair, but they are all reduced to the quality of down hairs. The amazing naked cat – the Canadian Sphynx – lacks both guard and awn hairs and has only a soft fuzz of down hairs on its extremities.

So selective breeding has played havoc with the natural coat of the cat, producing types of animal that would not all thrive in the wild today. Some would suffer from the cold, others from the heat, and still others would become badly matted and tangled without their daily grooming. Fortunately for these pedigree breeds there are usually plenty of human slaves around to tend to their needs and comforts and, should the worst happen and the animals be forced to fend for themselves as strays, changes would soon take place. They themselves might suffer from the climate, but if they managed to survive and interbreed the chances are that in a few generations their offspring would have reverted to wild-type coats once again, as a result of the inevitable mixing that would occur among the stray cat colonies.

Why does a cat lick its face when it is not dirty?

A quick flick of the tongue over the lips is one of the tell-tale signs that a cat is becoming agitated, while at the same time being fascinated or puzzled by something. Keeping an eye firmly fixed on the source of its agitation, the cat gives the impression that it has suddenly and inexplicably developed an urgent need to clean its nose or the fur around its mouth. But there is no dirt there. The cleaning is not functional and it does not follow the usual pattern seen after feeding or during a normal grooming session. The licks are short and sharp – rapid sweeps of the tongue that do not develop in the usual way into proper washing actions. They are the cat's equivalent of a man scratching his head when perplexed or irritated.

Reactions of this type are called 'displacement activities'. They occur when the cat is thrown into a state of conflict. Something which upsets it but at the same time arouses its curiosity, will simultaneously repel and attract the animal. There it sits, wanting to leave and wanting to stay. It stares at the irritant and, unable to resolve its conflict, shows its state of agitation by performing some trivial, abbreviated action – anything to break the stalemate in which it finds itself. Different species respond in different ways. Some animals nibble their paws, others scratch behind an ear with a hind leg. Birds wipe their beaks on a branch. Chimpanzees scratch their arms or their chins. But for felines the tongue-swipe is the favourite action.

There is a harmless way in which this can be tested. Cats do not like vibrating noises with a high pitch, but they are intrigued by what makes such sounds. A coin rubbed back and forth along the teeth of an ordinary hair-comb produces such a noise. Almost every cat, when hearing the brrrrrr sound produced by this action, stares at the comb in your hand and then, after a few seconds, starts licking its lips. If the sound continues, the animal may eventually decide that it has had enough and it will get up and walk away. Amazingly, this works for fully grown lions just as well as for small tabbies. Sometimes the lip-licking gives way to a violent sneeze, sometimes to a wide yawn. These actions appear to be alternative feline 'displacement activities', but they are less common than the lip-licking.

Why a cat is so irritated by a vibration sound is something of a mystery unless, during the course of feline evolution, it has come to represent a noxious animal of some kind – something unsuitable to attack as a prey. An obvious example that comes to mind here is the rattlesnake's rattle. Cats perhaps have an automatic alarm response to such animals and this may account for the fact that they are upset and yet intrigued at the same time.

How large is a cat's territory?

The wild counterpart of the domestic cat has a huge territory, with males patrolling up to 175 acres. Domestic cats which have gone wild and are living in remote areas where there is unlimited space also cover impressively large areas. Typical farm cats use nearly as much space, the males ranging over 150 acres. Female farm cats are more modest, using only about fifteen acres on average. In cities, towns and suburbs, the cat population becomes almost as overcrowded as that of the human citizens. The territories of urban cats shrink to a mere fraction of the home range enjoyed by their country cousins. It has been estimated that cats living rough in London, for example, enjoy only about one-fifth of an acre each. Pampered pet cats living in their owners' houses may be even more restricted, depending on the size of the gardens attached to the houses. The maximum density recorded is one pet cat per one-fiftieth of an acre.

This degree of variation in the size of feline territories shows just how flexible the cat can be. Like people, it can adjust to a massive shrinkage of its home ground without undue suffering. From the above figures it is easy to calculate that 8,750 crowded pet cats could be fitted into the territory of one wild cat living in a remote part of the world. The fact that the social life of the crowded cats does not become chaotic and vicious is a testimony to the social tolerance of cats. In a way this is surprising, because people often speak of the sociability of dogs, but stress that cats are much more solitary and unsociable. They may be so by choice, but given the challenge of living whisker-by-tail with other cats, they manage remarkably well.

They achieve this high-density success in a number of ways. The most important factor is the provision of food by their owners. This removes the need for lengthy daily hunting trips. It may not remove the urge to set off on such trips – a well-fed cat remains a hunting cat – but it does reduce the determination born of an empty stomach. If they find themselves invading neighbouring territories, they can give up the hunt without starving. If restricting their hunting activities to their own cramped home ranges makes them inefficient prey-catchers, it might prove frustrating, but it does not lead to starvation and death. It has been demonstrated that the more food the cats are given by their owners, the smaller their urban territories become.

Another factor helping them is the way in which their human owners divide up their own territories – with fences and hedges and walls to demarcate their gardens. These provide natural boundary-lines that are easy to recognize and defend. In addition there is a permissible degree of overlapping in feline territories. Female cats often have special areas where several of their home ranges overlap and where they can meet on neutral ground. The males – whose territories are always about ten times the size of those of the females, regardless of how great or small the crowding – show much more overlap. Each male will roam about on an area that takes in several female territories, enabling him to keep a permanent check on which particular queen (female) is on heat at any particular moment.

The overlapping is permitted because the cats are usually able to avoid one another as they patrol the landmarks in their patch of land. If, by accident, two of them do happen to meet up unexpectedly, they may threaten one another or simply keep out of one another's way, watching each other's movements and waiting their turn to visit a particular zone of the territory.

The numbers of pet cats are, of course, controlled by their owners, with the neutering of adults, destruction of unwanted litters and the selling or giving away of surplus kittens. But how does the territorial arrangement of feral cats survive the inevitable production of offspring? One detailed study of dockland cats at a large port revealed that in an area of 210 acres there were ninety-five cats. Each year they produced about 400 kittens between them. This is a high figure of about ten per female, which must mean that on average each queen gave birth to two litters. In theory this would mean a fivefold increase in the population each year. In practice it was found that the population remained remarkably stable from one year to the next. The cats had established an appropriate territory size for the feral, dockland world in which they lived, and then kept to it. Closer investigation revealed that only one in eight of the kittens survived to become adults. These fifty additions to the population each year were balanced by fifty deaths among the older cats. The main cause of death here (as with most urban cat populations) was the fatal road accident.

How sociable are cats?

The cat is often characterized as a solitary, selfish animal, walking alone and coming together with other cats only to fight or mate. When cats are living wild, with plenty of space, it is true that they do fit this picture reasonably well, but they are capable of changing their ways when they become more crowded. Living in cities and towns, and in the homes of their human owners, cats show a remarkable and unexpected degree of sociability.

Anyone doubting this must remember that, to a pet cat, we ourselves are giant cats. The fact that domestic cats will share a home with a human family is, in itself, proof of their social flexibility. But this is only part of the story. There are many other ways in which cats demonstrate co-operation, mutual aid, and tolerance. This is particularly noticeable when a female is having kittens. Other females have been known to act as midwives, helping to chew through the umbilical cords and clean up the new-born offspring. Later they may offer a babysitting service, bring food for the new mother, and occasionally feed young from other litters as well as their own. Even males sometimes show a little paternal feeling, cleaning kittens and playing with them.

These are not usual activities, but despite the fact that they are uncommon occurrences, they do reveal that the cat is capable, under special circumstances, of behaving in a less selfish way than we might expect.

Territorial behaviour also involves some degree of restraint and sharing. Cats do their best to avoid one another, and often use the same ranges at different times as a way of reducing conflict. In addition there are special no-cat's-land areas where social 'clubs' can develop. These are parts of the environment where, for some reason, cats call a general truce and come together without too much fighting. This is common with feral city cats, where there may be a special feeding site. If humans throw food for them there, they may gather quite peacefully to share it. Close proximity is tolerated in a way that would be unthinkable in the 'home base' regions of these cats.

Considering these facts, some authorities have gone so far as to say that cats are truly gregarious and that their society is more co-operative than that of dogs, but this is romantic exaggeration. The truth is that, where social life is concerned, cats are opportunists. They can take it or leave it. Dogs, on the other hand, can never leave it. A solitary dog is a wretched creature. A solitary cat is, if anything, relieved to be left in peace.

If this is so, then how can we explain the mutual aid examples given above? Some are due to the fact that we have turned domestic cats into overgrown kittens. By continuing to feed them and care for them we prolong their juvenile qualities into their adult lives. Like Peter Pan, they never grow up mentally, even though they become mature adults physically. Kittens are playful and friendly with their litter-mates and with their mothers, so they are used to acting together in a small group. This quality can be added to later adult activities, making them less competitive and less solitary. Secondly, those cats living wild in cities, where there is little space, adapt to their shrunken territories out of necessity, rather than by preference.

Some animals can live only in close-knit social groups. Others can tolerate only a completely solitary existence. The cat's flexibility means that it can accept either mode of living, and it is this that has been a key factor in its long success story since it was first domesticated thousands of years ago.

Why do cats keep crying to be let out and then cry to be let in again?

Cats hate doors. Doors simply do not register in the evolutionary story of the cat family. They constantly block patrolling activities and prevent cats from exploring their home range and then returning to their central, secure base at will. Humans often do not understand that a cat needs to make only a brief survey of its territory before returning with all the necessary information about the activities of other cats in the vicinity. It likes to make these tours of inspection at frequent intervals, but does not want to stay outside for very long, unless there has been some special and unexpected change in the condition of the local feline population.

The result of this is an apparent perversity on the part of pet cats. When they are in they want to go out, and when they are out they want to come in. If their owner does not have a small cat-flap on the back door of the house, there will be a regular demand for attention, to assist the cat in its rhythmic territorial supervision. Part of the reason why this repeated checking of the outside world is so important is because of the time-clock message system of the scent-marks. Each time a cat rubs against a landmark in its territory or sprays urine on it, it leaves a personal scent which immediately starts to lose its power. This decline is at a steady rate and can be used by cats to determine how long it has been since the scent-marker rubbed or sprayed. The repeated visits by a cat to inspect its territory are motivated by a need to reactivate its fading scent signals. Once this has been done, comfort and security beckons again, and the anxious feline face appears for the umpteenth time at the window.

What does a cat signal with its ears?

Unlike humans, felines have very expressive ears. They not only change direction as the cat listens to sounds coming from different sources, but they also adopt special postures that reflect the emotional mood of the animals.

There are five basic ear signals, related to the following moods: relaxed, alert, agitated, defensive and aggressive.

In the relaxed cat the apertures of the ears point forward and slightly outward, as the animal quietly listens for interesting sounds over a wide range.

When the resting cat stirs itself and focuses on some exciting detail in its surroundings, the ear position changes into the 'alert mode'. As it stares at the point of interest, its ears become fully erect and rotate so that their apertures point directly forward. The ears are kept pricked in this way as long as the cat remains gazing straight ahead. The only variation occurs if there is a sudden noise away to the side of the animal, in which case an ear may be permitted a brief rotation in that direction without a shift of gaze.

An agitated cat, suffering from a state of conflict, frustration or apprehension, often displays a nervous twitching of the ears. In some species of wild cats this response has been made highly conspicuous by the evolution of long ear-tufts, but the domestic cat lacks this refinement and the ear-twitching itself is less common. Slight tufting does occur in some breeds, especially the Abyssinian where there is a small dark hairy point to the ear, but compared with the huge ear-tufts of a species such as the Caracal Lynx, this is a very modest development.

A defensive cat displays fully flattened ears. They are pressed tightly against the head as a way of protecting them during fights. The torn and tattered ears of battling tom-cats are a vivid testimony to the need to hide this delicate part of the anatomy as much as possible when the claws are out. The effect of flattening the ears to the sides of the head is to make them almost invisible when the animal is viewed from the front and to give its head a more rounded outline. There is one strange breed of cat called the Scottish Fold which has permanently flattened ears, giving it a continually defensive look, regardless of its true mood. What effect this has on its social life is hard to imagine.

An aggressive cat which is hostile without being particularly frightened has its own special ear posture. Here, the ears are rotated but not fully flattened. The *backs* of the ears become visible from the front, and this is the most dangerous ear signal any cat can transmit. In origin, this ear posture is half-way between alert and defensive – in other words, half-way between pricked forward and flattened backward. In effect, it is a 'ready for trouble' position. The animal is saying, 'I am ready to attack, but you don't frighten me enough to flatten my ears protectively.' The reason why this involves showing off the backs of the ears is because they must be rotated backwards before they can be fully flattened. So the rotated ears are in a 'ready-to-be-flattened' posture, should the aggressive cat's opponent dare to retaliate.

The aggressive ear posture has led to some attractive ear-markings in a number of wild cat species, especially the tiger, which has a huge white spot ringed with black on the back of each ear. When a tiger is angry, there is no doubt at all about its mood, as the pair of vivid white spots rotates into view. Again, domestic cats lack these special markings.

How do cats fight?

Under wild conditions cat-fights are a rarity because there is plenty of space, but in the more crowded urban and suburban areas feline territories become squashed together and frequently overlap. This means that a great deal of squabbling and serious physical duelling occurs, especially between rival tom-cats. Occasionally there are even killings or deaths resulting from battle injuries.

The primary objective of an attacking cat is to deliver a fatal neck-bite to its rival, employing much the same technique as when killing a prey. Because its opponent is of roughly the same size and strength, this lethal bite is hardly ever delivered. Indeed, the most craven and cowardly of rivals will defend itself to some extent, and a primary neck-bite is almost impossible to achieve.

The point to remember here is that even the most savage and dominant individual, as he goes in to the attack, is fearful of the consequences of the 'last-ditch-stand' by his terrorized underling. Driven into a corner, the weakling will try anything, lashing out with sharp claws and possibly injuring the dominant cat in a way that may pose a serious threat to his future hunting success and therefore his very survival. So even an out-and-out attacker shows fear mixed with his aggression, when the final crunch of physical contact arrives.

A typical sequence goes as follows: the dominant animal spots a rival and approaches it, adopting a highly characteristic threat posture, walking tall on fully stretched legs so that it suddenly appears bigger than usual. This effect is increased by the erecting of the hairs along its back. Because the crest is greater towards the rear end of the animal, the line of its back slopes up towards the tail. This gives the attacking cat a silhouette which is the exact opposite of the crouching shape of the weaker rival, whose rear end is held low on the ground.

With the backs of his ears showing and a great deal of howling, growling and gurgling, the attacker advances in slow motion, watching for any sudden reaction from his cringing enemy. The noises made are startlingly hostile and it is hard to understand how anything so totally aggressive can ever have been misnamed the tom-cat's 'love song'. One can only wonder at the love-life of the people who gave it this name. Needless to say, it has nothing whatever to do with true cat courtship.

As the attacking cat comes very near its rival, it performs a strange but highly characteristic head-twisting action. At a distance of about three feet it raises its head up slightly and then tilts it over to one side, all the time fixating the enemy with its eyes. Then the attacker takes a slow step forward and tilts its head the other way. This may be repeated several times and appears to be a threat of the neck-bite to come, the head twisting into the biting position as much as to say 'this is what you will get'. In other words, the attacker acts out the 'intention movement' of the assault typical of the species.

If two cats of equal status meet and threaten one another, a long period of deadlock may follow, with each animal performing exactly the same slow, hostile approach, as if displaying in front of a mirror. The nearer they get, the slower and shorter are their movements, until they become frozen in a prolonged stalemate which may last for many minutes. Throughout this they will continue to give vent to their caterwauling howls and moans, but neither side will be prepared to capitulate. Eventually they may separate from one another in incredibly slow motion. To increase their speed would be tantamount to admitting weakness and would lead to an immediate attack from the rival, so they must both withdraw with almost imperceptible movements to retain their status.

Should these threats and counter-threats collapse into a serious fight, the action begins with one of the adversaries making a lunging attempt at a neck-bite. When this happens the opponent instantly twists round and defends itself with its own jaws, while at the same time striking out with its front feet, clinging on with its forepaws and then kicking wildly with its powerful back feet. This is the point at which the 'fur flies' quite literally, and the growling gives way suddenly to yowls and screams as the two animals roll and writhe around, biting, clawing and kicking.

This phase does not last long. It is too intense. The rivals quickly pull apart and resume the threat displays, staring at one another and growling throatily once again. The assault is then repeated, perhaps several times, until one of them finally gives up and remains lying on the ground with its ears fully flattened. At this point the victor performs another highly characteristic display. It turns at right angles to the loser and, with great concentration, starts to sniff the ground, as though at that very moment there is an irresistibly delicious odour deposited there. The animal concentrates so hard on this sniffing that, were it not a regular feature of all fights, it would have the appearance of a genuine odour-check. But it is now only a ritual act, a victory display which signals to the cowering rival that its submission and capitulation have been accepted and that the battle is over. After the ceremonial sniffing the victor saunters slowly off and then, after a short while, the vanquished animal slinks away to safety.

Not all fights are conducted at such high intensity. Milder disputes are settled by 'paw-scrapping' in which the rivals swipe out at one another with extended claws. Slashing at their rival's head in this way, they may be able to settle their disagreement without the full ritual battle and all-in wrestling described above.

Why does a cat arch its back when it sees a strange dog?

If a cat feels threatened by a large dog, it pulls itself up on fully stretched legs and at the same time arches its back in the shape of an inverted U. The function of this display is clearly to make the cat look as big as possible, in an attempt to convince the dog that it is confronting a daunting opponent. To understand the origin of the display it is necessary to look at what happens when cats are threatening one another. If one cat is intensely hostile towards another and feels little fear, it approaches on stiffly stretched legs and with a straight back. If its rival is extremely frightened and feels no hostility, it arches its back and crouches low on the ground. In the case of the cat approached by a dog, there is both intense aggression *and* intense fear. It is this conflicting, double mood that gives rise to the special display. The cat borrows the most conspicuous element of its anger reaction – the stiff legs – and the most conspicuous element of its fear reaction – the arched back – and combines them to produce an 'enlarged cat' display. If it had borrowed the other elements – the straight back of anger and the low crouch of fear – the result would have been far from impressive.

Aiding its 'transformation display' is the fact that the animal, while stretching its legs and arching its back, also erects its fur and stands broadside-on to the dog. Together these four elements make up a compound display of maximum size increase. Even if the cat retreats a little, or advances towards the dog, it carefully keeps its broadside-on position, spreading its body in front of the dog like a bullfighter's cloak.

During the arched-back display the cat hisses ominously, like a snake, but this hissing turns to growling if it risks an attack. Then, when it actually lashes out at the dog, it adds an explosive 'spit' to its display. Experienced cats soon learn that the best policy when faced with a hostile dog is to go into the attack rather than run away, but it takes some nerve to do this when the dog is several times the cat's weight. The alternative of 'running for it' is much riskier, however, because once the cat is fleeing it triggers off the dog's hunting urges. To a dog a 'fleeing object' means only one thing — food — and it is hard to shift the canine hunting mood once it has been aroused. Even if the fleeing cat halts and makes a brave stand, it has little hope, because the dog's blood is up and it goes straight for the kill, arched back or no arched back. But if the cat makes a stand right from the first moment of the encounter with the dog, it has a good chance of defeating the larger animal, simply because by attacking it, the cat gives off none of the usual 'prey signals'. The dog, with sharp claws slashing at its sensitive nose, is much more likely then to beat a dignified retreat, and leave the hissing fury to its own devices. So, where dogs are concerned, the bolder the cat, the safer it is.

Why does a cat hiss?

It seems likely that the similarity between the hiss of a cat and that of a snake is not accidental. It has been claimed that the feline hiss is a case of protective mimicry. In other words, the cat imitates the snake to give an enemy the impression that it too is venomous and dangerous.

The quality of the hissing is certainly very similar. A threatened cat, faced with a dog or some other predator, produces a sound that is almost identical to that of an angry snake in a similar situation. Predators have great respect for venomous snakes, with good reason, and often pause long enough for the snake to escape. This hesitation is usually the result of an inborn reaction. The attacker does not have to learn to avoid snakes. Learning would not be much use in such a context, as the first lesson would also be the last. If a cornered cat is capable of causing alarm in an attacker by triggering off this instinctive fear of snakes, then it obviously has a great advantage, and this is probably the true explanation of the way in which the feline hiss has evolved.

Supporting this idea is the fact that cats often add spitting to hissing. Spitting is another way in which threatened snakes react. Also, the cornered cat may twitch or thrash its tail in a special way, reminiscent of the movements of a snake that is working itself up to strike or flee.

Finally, it has been pointed out that when a tabby cat (with markings similar to the wild type, or ancestral cat) lies sleeping, curled up tightly on a tree-stump or rock, its coloration and its rounded shape make it look uncannily like a coiled snake. As long ago as the nineteenth century it was suggested that the pattern of markings on a tabby cat are not direct, simple camouflage, but rather are imitations of the camouflage markings of a snake. A killer, such as an eagle, seeing a sleeping cat might, as a result of this resemblance, think twice before attacking.

Why does a cat wag its tail?

Most people imagine that if a cat wags its tail it must be angry, but this is only a partial truth. The real answer is that the cat is in a state of conflict. It wants to do two things at once, but each impulse blocks the other. For example, if a cat cries to be let out at night and the door is opened to reveal a downpour of heavy rain, the animal's tail may start to wag. If it rushes out into the night and stands there defiantly for a moment, getting drenched, its tail wags even more furiously. Then it makes a decision and either rushes back in to the comforting shelter of the house, or bravely sets off to patrol its territory, despite the weather conditions. As soon as it has resolved its conflict, one way or the other, its tail immediately stops wagging.

In such a case it is inappropriate to describe the mood as one of anger. Anger implies a frustrated urge to attack, but the cat in the rainstorm is not aggressive. What is being frustrated there is the urge to explore, which in turn is frustrating the powerful feline desire to keep snug and dry. When the two urges momentarily balance one another, the cat can obey neither. Pulled in two different directions at once, it stands still and wags its tail. Any two opposing urges would produce the same reaction, and only when one of these was the urge to attack – frustrated by fear or some other competing mood – could we say that the cat was wagging its tail because it was angry.

If tail-wagging in cats represents a state of acute conflict, how did such a movement originate? To understand this, watch a cat trying to balance on a narrow ledge. If it feels itself tipping over, its tail quickly swings sideways, acting as a balancing organ. If you hold a cat on your lap and then tip it slightly to the left and then to the right, alternating these movements, you can see its tail swinging rhythmically from side to side as if it is wagging it in slow motion. This is how the tail-wagging movement used in mood-conflicts began. As the two competing urges pulled the cat in opposite directions, the tail responded as if the animal's body were being tipped over first one way and then the other. During evolution this lashing of the tail from side to side became a useful signal in the body language of cats and was greatly speeded up in a way that made it more conspicuous and instantly recognizable. Today it is so much faster and more rhythmic than any ordinary balancing movement of the tail that it is easy to tell at a glance that the conflict the animal is experiencing is emotional rather than purely physical.

How many tail-signals does a cat make?

In addition to the familiar tail-wagging of a cat in a conflict, there are a number of other tail-signals that indicate the changing moods of the pet feline as it goes about its business. Each tail movement or posture tells us (and other cats) something about the animal's emotional condition and it is possible to draw up a 'de-coding key', as follows:

Tail curves gently down and then up again at the tip
    This is the relaxed cat, at peace with the world.
Tail raised slightly and softly curved
    The cat is becoming interested in something.
Tail held erect but with the tip tilted over
    The cat is very interested and is in a friendly, greeting mood, but with slight reservations.
Tail fully erect with the tip stiffly vertical
    An intense greeting display with no reservations. In adult cats, this posture is 'borrowed' from the action of a kitten greeting its mother. The kitten's signal is an invitation to the mother cat to inspect its rear end, so there is an element of subordination in this display, as there is in most greeting ceremonies.
Tail lowered fully and possibly even tucked between the hind legs
    This is the signal of a defeated or totally submissive cat that wishes to stress its lowly social status.
Tail lowered and fluffed out
    The cat is indicating active fear.
Tail swished violently from side to side
    This is the conflict signal of tail-wagging, in its most angry version. If the tail swings very vigorously from side to side it usually means that the animal is about to attack, if it can summon up that last ounce of aggression.
Tail held still, but with tip twitching
    This is the version of tail-wagging that indicates only mild irritation. But if the tip-twitching becomes more power-

ful, then it can act as a clue that a swipe from a bad-tempered paw is imminent.

Tail held erect with its whole length quivered

This gentle quivering action is often seen after a cat has been greeted by its owner. It is the same action that is observed when urine-spraying is taking place out of doors, but in this case there is no urine produced. Whether some slight, invisible scent is expelled is not clear, but the gesture appears to have the meaning of a friendly 'personal identification' as if the cat is saying, 'Yes, this is *me*!'

Tail held to one side

This is the sexual invitation signal of the female cat on heat. When she is ready to be mounted by the male she conspicuously moves her tail over to one side. When he sees this, the tom-cat knows he can mount her without being attacked.

Tail held straight and fully bristled

This is the signal of an aggressive cat.

Tail arched and bristled

This is the signal of a defensive cat, but one that may attack if provoked further. The bristling of the fur makes the animal look bigger, a 'transformation display' that may deter the enemy if the defensive cat is lucky.

When do cats become sexually mature?

This occurs when they are nearly a year old, but there is a great deal of variation. For toms, the youngest recorded age for sexual maturity is six months, but this is abnormal. Eight months is also rather precocious, and the typical male does not become sexually active until he is between eleven and twelve months of age. For toms living rough, it may be considerably longer – more like fifteen to eighteen months, probably because they are given less chance in the competition with older males.

For females, the period can be relatively short, six to eight months being usual, but very young females only three to five months old have been known to come into sexual condition. This early start seems to be caused by the unnatural circumstances of domestication. For a wild cat, ten months is more usual.

The European Wild Cat, for instance, starts its breeding season in March. There is a gestation period of sixty-three days and then the kittens appear in May. By late autumn they strike off on their own and, if they survive the winter, they will themselves start to breed the following March when they are about ten months old, producing their own litters when they are a year old. For these wild cats there is only one season a year, so young toms may have to be patient and wait for the following season before they go into action.

This wild cycle is obviously geared to the changing seasons and the varying food supply, but for the pampered pet there are no such problems. With its hunting ears finely tuned to the metallic sound of a can-opener and with the central-heating humming gently in the background, the luxuriating house cat has little to fear from the annual cycle of nature. As a result, its breeding sequences are less rigid than its wild counterpart. It may breed as early as the second half of January, producing a litter by the beginning of April. Two months later, with its kittens weaned and despatched to new homes, it may well start off again with another breeding sequence, producing a second litter in the late summer. With this loss of a simple annual rhythm, there is a whole scatter of ages among young domestic cats, leading to the variations in the stages at which they become sexually active.

Cases have been reported of wild cats producing a second litter in August, but it is suspected that this only occurs where there has been interbreeding between the wild animals and feral domestic cats.

How fast do cats breed?

Without restraint a cat population can increase at a startling rate. This is because female cats are excellent mothers, and because domestication has led to a possible tripling of the number of litters and to an increase in litter size. European Wild Cats, with their single litter each year, have an average of two to four kittens, but domestic cats may produce an average of four to five kittens in each of their three annual litters.

A simple calculation, starting with a single breeding pair of domestic cats, and allowing for a total of fourteen kittens in each three-litter year, reveals that in five years' time there will be a total of 65,536 cats. This assumes that all survive, that males and females are born in equal numbers and that they all start breeding when they are a year old. In reality, the females might start a little younger, so the figure could be higher. But against this is the obvious fact that many would perish from disease or accident.

This paints a grim picture for the aspiring house mouse, a nightmare world of wall-to-wall cats. But it never materializes because there are enough responsible human owners to ensure that breeding restraints *are* applied to their pet cats, to keep the numbers under control. Neutering of both males and females is now commonplace and it is estimated that more than 90 per cent of all toms have suffered the operation. Females that are allowed to breed may have their litter size reduced to one or two, the unfortunate kittens being painlessly killed by the local vet. In some areas there are fairly ruthless extermination programmes for feral and stray cats, and in certain countries there have even been oral contraceptive projects, with the stray cat population given food laced with 'the pill'. Israel, for example, claims to prevent about 20,000 kittens a year by using this technique.

Despite these attempts, however, there are still well over a million feral and stray cats in Great Britain at the present time. It has been estimated that there are as many as half a million in the London region alone. In addition there are between four and five million pet cats, making a massive feline population of roughly one cat per ten humans.

How do cats perform their courtship?

Cats spend a great deal of time building up to the mating act, and their prolonged 'orgies' and promiscuity have given them a reputation for lasciviousness and lust, over the centuries. This is not because the mating act itself is lengthy or particularly erotic in form. In fact, the whole process of copulation rarely exceeds ten seconds and is often briefer than that. What gives the felines their reputation for lechery is the superficial resemblance between their sexual gatherings and a Hell's Angels gang-bang. There is a female, spitting and cursing and swiping out at the males one moment and writhing around on the ground the next. And there is a whole circle of males, all growling and howling and snarling at one another as they take it in turns (apparently) to rape the female.

The truth is slightly different. Admittedly, the process may take hours, even days, of non-stop sexual activity, but it is the female who is very much in charge of what is happening. It is she who calls the tune, not the males.

It begins when the female comes on heat and starts calling to the males. They also respond to her special sexual odours and are attracted from all around. The male on whose territory she has chosen to make her displays is initially strongly favoured, because other males from neighbouring territories will be scared at invading his ground. But a female on heat is more than they can resist, so they take the risk. This leads to a great deal of male-to-male squabbling (and accounts for most of the noise – which is why the caterwauling and howling is thought of, mistakenly, as sexual, when in reality it is purely aggressive). But the focus of interest is the female and this helps to damp down the male-to-male fighting and permits the gathering of a whole circle of males around her.

She displays to them with purring and crooning and rolling on the ground, rubbing herself and writhing in a manner that fascinates the male eyes fixating upon her. Eventually one of the males, probably the territory owner himself, will approach her and sit close to her. For his pains, he is attacked with blows from her sharp-clawed forepaws. She spits and growls at him and he retreats. Any male approaching her too soon is seen off in this way. She is the mistress of the situation and it is she who will eventually choose which male may approach her more intimately. The male who succeeds in this may or may not be the dominant tom present. That is up to her. But certain male strategies do help the toms to succeed. The most important one is to advance towards her only when she is looking the other way. As soon as she turns in his direction the male freezes — like a child playing the party game called 'statues'. She only attacks when she sees the actual advance itself, not the immobile body that has somehow, by magic, come a little nearer. In this way a tom with finesse can get quite close. He offers her a strange little chirping noise and, if she gives up spitting and hissing at him, he will eventually risk a contact approach. He starts by grabbing the scruff of her neck in his jaws and then carefully mounts her. If she is ready to copulate she flattens the front of her body and raises her rump up into the air, twisting her tail to one side. This is the posture called 'lordosis' and is the final invitation signal to the male, permitting him to copulate.

As time goes on, the 'orgy' changes its style. The males become satiated and are less and less interested in the female. She, on the other hand, seems to become more and more lustful. Having worked her way through one male after another at comparatively short intervals for perhaps several days, one might imagine that she too would be satiated, but this is not so. As long as her peak period of heat persists she will want to be mated, and the toms now have to be encouraged. Instead of playing hard to get, she now has to work on the males to arouse their interest. She does this with a great deal of crooning, rubbing and especially writhing on the ground. The males still sit around watching her, and from time to time manage to muster enough enthusiasm to mount her once more. Eventually it is all over and the chances of a female cat returning home unfertilized after such an event are utterly remote.

Why does the tom grab the female by the scruff of the neck when mating?

At first sight this appears to be a piece of macho brutality, in similar vein to the cartoon caveman grabbing his mate by her hair and dragging her off to his cave. Nothing could be further from the truth. In sexual matters it is the female, not the male, who is dominant, where cats are concerned. Toms may fight savagely among themselves, but when they are sexually aroused and attempting to mate with the queen they are far from bossy. It is the female who swipes out and beats the toms. The bite on the back of her neck may look savage, but in reality it is a desperate ploy on the part of the male to protect himself from further assault. This protection is of a special kind. It is not a matter of forcibly holding down the female so that she cannot twist round and attack him. She is too strong for that. Instead it is a 'behaviour trick' played by the male. All cats, whether male or female, retain a peculiar response to being grabbed firmly by the scruff of the neck, dating back to their kitten days. Kittens have an automatic reaction to being held in this way by their mother. She uses it when it is necessary to transport the kittens from an unsafe to a safe place. It is crucially important that the kittens do not struggle on such occasions, where their very lives may be at stake. So felines have evolved a 'freeze' reaction to being taken by the scruff of the neck − a response which demands that they stay quite still and do not struggle. This helps the mother in her difficult task of moving the litter to safety. When they grow up, cats never quite lose this response, as you can prove to yourself by holding an adult pet cat firmly by the skin of its neck. It immediately stops moving and will remain immobile in your grasp for some time before becoming restless. If you grasp it firmly on some other part of its body the restlessness is much quicker to occur, if not instantaneous. This 'immobilization reaction' is the trick the toms apply to their potentially savage females. The females are so claw-happy that the toms badly need such a device. As long as they hang on with their teeth, they have a good chance that the females will be helplessly transformed into 'kittens lying still in their mother's jaws'. Without such a behaviour trick the tom would return home with even more scars than usual.

Why does the female scream during the mating act?

As the tom finishes the brief act of copulation, which lasts only a few seconds, his female twists round and attacks him, swiping out savagely with her claws and screaming abuse at him. As he withdraws his penis and dismounts he has to move swiftly, or she is liable to injure him. The reason for her savage reaction to him at this point is easy to understand if you examine photographs of his penis taken under the microscope. Unlike the smooth penis of so many other mammals, the cat's organ is covered in short, sharp spines, all pointing away from the tip. This means that the penis can be inserted easily enough, but when it is withdrawn it brutally rakes the walls of the female's vagina. This causes her a spasm of intense pain and it is this to which she reacts with such screaming anger. The attacked male, of course, has no choice in the matter. He cannot adjust the spines, even if he wishes to do so. They are fixed and, what is more, the more sexually virile the male, the bigger the spines. So the sexiest male causes the female the most pain.

This may sound like a bizarre sado-masochistic development in feline sex, but there is a special biological reason for it. Human females who fail to become pregnant ovulate at regular intervals, regardless of whether they have mated with a male or not. Human virgins, for example, ovulate month after month, but this is not the case with cats. A virgin cat would not ovulate at all. Cats *only* ovulate after they have been mated by a male. It takes a little while – about twenty-five to thirty hours – but this does not matter because the intense period of heat lasts at least three days, so she is still actively copulating when ovulation occurs. The trigger that sets off the ovulation is the intense pain and shock the queen feels when her first suitor withdraws his spiny penis. This violent moment acts like the firing of a starting pistol which sets her reproductive hormonal system in operation.

In a way, it is not far from the truth to call a female cat on heat 'masochistic' because, within about thirty minutes of having been hurt by the first male penis, she is actively interested in sex again and ready to be mated once more, with a repeat performance of the scream-and-swipe reaction. Considering how much the spiny penis must have hurt her, it is clear that in a sexual context there is one kind of pain which does not produce the usual negative response.

How long do cats continue to breed?

The short answer is: for a very long time. Tom-cats have been known to produce offspring at the advanced feline age of sixteen years. This is equivalent to a human male becoming a father in his late seventies.

Female cats have been known to give birth when twelve years old. For a human female this would be like having a baby in her mid-sixties. The oldest known woman to give birth was fifty-seven years old, which is the equivalent of only nine feline years. The more typical female, experiencing the menopause at fifty-one, would be the equivalent of a female cat of only seven years. This means that cats remain fertile longer than we do, in relative terms.

Not to exaggerate the cat's breeding abilities, it must be recorded that from the age of eight until twelve years there is a gradual decline in the number of kittens produced in each litter, so the reproductive apparatus is beginning to show signs of slowing down at this stage, and only the strongest and healthiest of moggies can stay the full course. Pedigree cats, because they lack 'hybrid vigour', are not so long-lasting.

For those who like to make comparisons between their pets' ages and their own, bearing in mind that the figures are only a rough guide, the following table may be of interest. The ages are given in years:

Your cat's age	Your own age
1	15
2	25
4	40
7	50
10	60
15	75
20	105
30	120

Are there contraceptives for cats?

Yes, there are. Because cats, like people, are breeding too fast there has been a concerted effort to provide them with a contraceptive pill similar to the one we use. A few years ago this was extremely popular, but since the initial boom in feline birth pills there has been a decline in interest in favour of the more drastic method of neutering. This is partly because, in early field trials with stray cats, it was discovered that the animals were developing side-effects of an unpleasant kind, and partly because there is a less than 100 per cent efficiency. Since stray cat populations can only be dosed with the birth pill via their food, there is always the chance that certain cats will avoid the treated meals provided for them and scavenge or catch prey for themselves. Such individuals will then continue to breed, and the numbers will not sink as rapidly as the population controllers would like.

There are several kinds of contraceptive pills available, and they act in slightly different ways. The *progestogens* have the same effect on the cat's body as the natural pregnancy hormone, progesterone. They give the female cat a false pregnancy, complete with all the usual accompanying symptoms, such as increased appetite and increased weight. They can be administered either as simple tablets or as a special, long-acting injection. But in both instances there are dangers of infection and for this reason other methods have since been tried.

A modified version of this treatment employing weaker progestogens has been tested recently and there is now a much safer pill of this type available. Called *proligesterone*, its side-effects appear to be much less damaging.

A different approach is to inhibit the hormone which starts off the female sexual cycle. This hormone, called gonadotrophin, can be suppressed by certain drugs that stop the oestrus cycle without causing serious side-effects. This is a new method and is being developed further with some optimism.

A non-chemical method is also possible, but requires skilful, expert handling. This involves stimulating a female cat that is on heat with a glass rod, so that her body is fooled into reacting as if she has been mated by a tom-cat. Because it is the mating act in cats that induces ovulation, it is possible in this way to start the female cat's ovulation as if she is carrying male sperm. Because she is not, the eggs she sheds will be wasted and contraception will have been achieved. Her sexual appetite will pass and she will be quiet again until her next heat. As before, however, she will have to go through a phantom pregnancy as a result of this treatment.

All these methods require veterinary assistance and should not be attempted without professional help or approved prescriptions. There is no doubt, though, that within the next half-century we will see this type of biological control of feline populations perfected.

How does being neutered affect a cat's behaviour?

It has become increasingly common to remove the sex organs of both male and female domestic cats. In books written by vets on the subject of cat 'care' and cat 'health', it is now standard practice to refer to the spaying of females and the castration of males as minor, routine operations. 'Unless you are setting up a breeding stud, all pet toms should be castrated,' is a typical comment. What is not recorded is the tom-cats' attitude to this helpful form of health care. It is just possible that, given the choice, they might prefer the risk of an occasional torn ear or scratched nose to the certainty of a totally sexless adult life.

How has it come about that we are prepared to describe a serious physical mutilation as a trivial adjustment? Why are we so ready to treat as a minor operation an alteration which involves a major transformation in the cat's lifestyle and personality? The answer is that, for many of us, the cat has become a living toy rather than a real animal. We enjoy its company but we are not prepared to tolerate any inconvenience. So, if it is a male, we cut off its testicles; if it is a female we cut out its uterus and its ovaries; and if we are house-proud, we also cut out the claws of both sexes. Our sexless, clawless cats will now give us the perfect companionship we seek. They will not yowl, wander, mate or engage in sexual brawling; they will not spray their sexual scents; they will not tear precious fabrics; nor will they be able to hunt, kill or climb with any efficiency. In short, they will not be true cats, but they will undeniably make more convenient pets. In the end it all comes down to a matter of priorities. How much trouble are we prepared to go to for the privilege of sharing our lives with feline companions?

In order to persuade us to carry our cats off to the nearest surgery for 'altering', the many benefits of the operation are extolled at length. Apart from eliminating the tiresome sexual preoccupations of our pets, it will also make them more affectionate and more playful. Their average lifespan will be increased by two or three years. They will become generally more docile and less demanding. The surgeon's knife beckons invitingly.

Should we hesitate, there is one final threat to drive us on: the spectre of over-population. If we fail to remove our cats' sex organs, they will fill the world with thousands of unwanted kittens. If this were true it would be a powerful argument, but it is not. There is an alternative. It is easy to prevent cats breeding without de-sexing them. The most obvious way is to keep them indoors at times of sexual activity. This does, however, prove so difficult in most cases that it is not recommended. The better way to handle the situation is to render the cats infertile without actually neutering them. For the female this means cutting or tying her fallopian tubes. This prevents her from becoming pregnant but does not interfere with her love-life. For the male it means a vasectomy – the cutting or tying of the male's sperm ducts. This prevents him from supplying sperm to fertilize his females, but again it does nothing else. It does not make him docile or lazy, nor does it interfere with his sexual activities.

The existence of this contraceptive technique immediately puts cat owners on the spot. It prevents those who do neuter their pets from claiming that they are simply responsible citizens who wish to assist in the elimination of starving strays. The thousands of strays created by reckless, thoughtlessly uncontrolled cat-breeding could all have been prevented by the tube-tying techniques, without resorting to the full butchery of neutering. The owners are forced to admit honestly that they are having their cats' sex organs cut out for purely selfish reasons – to make them less restless, noisy and, in the case of toms, less smelly and belligerent in the mating seasons. If they face this honestly it is somehow less offensive than if they pretend that they are having their pets mutilated for 'the cat's own good'.

Why do cats sneer?

Every so often a cat can be seen to pause and then adopt a curious sneering expression, as if disgusted with something. When first observed, this reaction was in fact called an 'expression of disgust' and described as the cat 'turning up its nose' at an unpleasant smell, such as urine deposited by a rival cat.

This interpretation is now known to be an error. The truth is almost the complete opposite. When the cat makes this strange grimace, known technically as the *flehmen* response, it is in reality appreciating to the full a delicious fragrance. We know this because tests have proved that urine from female cats in strong sexual condition produces powerful grimacing in male cats, while urine from females not in sexual condition produces a much weaker reaction.

The response involves the following elements: the cat stops in its tracks, raises its head slightly, draws back its upper lip and opens its mouth a little. Inside the half-opened mouth it is sometimes possible to see the tongue flickering or licking the roof of the mouth. The cat sniffs and gives the impression of an almost trancelike concentration for a few moments. During this time it slows its breathing rate and may even hold its breath for several seconds, after sucking in air. All the time it stares in front of it as if in a kind of reverie.

If this behaviour were to be likened to a hungry man inhaling the enticing smells emanating from a busy kitchen, it would not be too far from the truth, but there is an important difference. For the cat is employing a sense organ that we sadly lack. The cat's sixth sense is to be found in a small structure situated in the roof of the mouth. It is a little tube opening into the mouth just behind the upper front teeth. Known as the vomero-nasal or Jacobsen's organ, it is about half an inch long and is highly sensitive to airborne chemicals. It can best be described as a taste-smell organ and is extremely important to cats when they are reading the odour-news deposited around their territories. During human evolution, when we became increasingly dominated by visual input to the brain, we lost the use of our Jacobsen's organs, of which only a tiny trace now remains, but for cats it is of great significance and explains the strange, snooty, gaping expression they adopt occasionally as they go about the social round.

Are there any gay cats?

Yes and no. Yes, cats do perform homosexual acts under certain circumstances, but no, they never prefer to mate with members of their own sex if members of the opposite sex are present. Homosexual acts are always second best for them and they are in no sense sexually disabled, as are certain human males or females who are incapable of being aroused sexually by the opposite sex.

If two male or two female cats find themselves together, sexually aroused but lacking suitable mates, one member of the pair may suddenly switch to the mating pattern of the 'wrong' sex. A female may mount another female, showing the masculine neck-bite and sometimes even making the typical male body-thrusts. Similarly, a male may crouch like a female and perform pseudo-female behaviour, being mounted by another tom.

This is simply an overspilling of sexual activity under conditions of extreme frustration, and the animals which have been seen to perform in this way have also been observed to copulate normally afterwards, if mates are provided. If the sexual thwarting becomes acute enough, some cats will even masturbate or attempt to mate with inanimate objects. One cat was observed to attempt copulation with a toy teddy bear, another with a toy rabbit.

In laboratory studies where male cats were kept isolated, awaiting mating tests with females, it was discovered that if one male was placed in a cage belonging to another tom it was often 'raped' by the cage owner. This was not a matter of individual strength but of territorial ownership. Even if the cage owner was a small, weaker cat and the introduced male was large and powerful, it was still the owner which mounted and the visitor which crouched in the female fashion. If the situation was reversed, then the mating roles were also reversed. It was always the territory owner who (literally) came out on top.

Is it possible to catch AIDS from cats?

No, it is not. But it has become difficult to convince those hypochondriacs who also happen to be cat owners that there is no danger. This is because certain newspapers thoughtlessly mentioned that some pet cats in California had been discovered to be suffering from the AIDS virus. These reports were made without checking the facts and without any consideration for the panic they could cause. Inevitably they led to the totally unnecessary deaths of many pet cats, as jittery owners took emergency steps to protect themselves from the dreaded twentieth-century plague.

Their fears were quite unfounded, because the so-called 'Feline AIDS' is caused by a different virus from the one that has attacked humans. True, it belongs to the same group of viruses, but within that group it is only distantly related. So, even if bitten or scratched by a cat that had somehow smeared its teeth or claws with infected blood from its sores, the human victim would still not be able to pick up the disease. There is no evidence from any source that 'Feline AIDS' can infect the human body.

Despite this, scaremongering journalists have managed to cause fear and anxiety among those cat-lovers who are of a 'nervous disposition'. Within hours of the reports appearing in the press, vets and cat sanctuaries were flooded with requests to have cats destroyed or to find them new homes. By the time that the truth had been publicized it was sadly too late for many unfortunate cats, and even now the ghost of the 'Feline AIDS' still stalks the world of cats and cat owners. For the felines themselves it is almost like a modern rebirth of the senseless persecution of the Medieval period, when they were accused of being servants of Satan. We can only hope that common sense will prevail more quickly than it did in the previous instance.

How does a female cat deal with her new-born kittens?

As the nine-week gestation period comes to an end the pregnant cat becomes restless, searching around for a suitable den or nest in which to deliver her kittens. She looks for somewhere quiet, private and dry. In a house, strange noises emanate from cupboards and other nooks and crannies as the cat tests out a variety of suitable sites. Suddenly, from being increasingly ravenous, her hunger vanishes and she refuses food, which means that the moment of birth is imminent – perhaps only a few hours away. At this point she disappears and settles down to the serious business of bringing a litter of kittens into the world.

Some cats hate interference at this stage and become upset by too much attention. Others – usually those that have never been given much privacy in the house – do not seem to care much one way or the other. The happy-go-lucky ones will co-operatively move into a specially prepared birth-box, with soft warm bedding provided and easy accessibility for a human midwife, should one be needed. Other cats stubbornly refuse the perfect nest-bed offered them and perversely disappear into the shoe-cupboard or some such dark, private place.

Giving birth is a lengthy process for the average cat. With a typical litter of, say, five kittens, and with a typical delay of, say, thirty minutes between the arrival of each one, the whole process lasts for two hours, after which both cat and kittens are quite exhausted. Some cats give birth much more quickly – one kitten per minute – but this is rare. Others may take as long as an hour between kittens – but this is also uncommon. The typical time delay of about half an hour is not an accident. It gives the mother long enough to attend to one kitten before the next arrives.

The attention she gives the new-born baby consists of three main phases. First, she breaks away the birth sac (the amniotic sac) which encases the kitten as it emerges into the world. She then pays special care to the cleaning of the nose and mouth of the new-born, enabling it to take its first breath. Once this crucial stage is over, she starts to clean up, biting through the umbilical cord and eating it, up to about one inch from the kitten's belly. The little stump she leaves alone, and this eventually dries out and finally drops off of its own accord. She then eats the afterbirth – the placenta – which provides her with valuable nourishment to see her through the long hours of total kitten-caring that now face her, during their first day of life. After this she licks the kitten all over, helping to dry its fur, and then she rests. Soon the next kitten will appear and the whole process will have to be repeated. If she grows tired, towards the end of an unusually large litter, the last one or two kittens may be ignored and left to die, but most female cats are amazingly good midwives and need no help from their human owners.

As the kittens recover from the trauma of birth, they start rooting around, searching for a nipple. The first feed they enjoy is vitally important because it helps to immunize them against disease. Before she produces her full-bodied nutritional milk, the mother provides a thin first-milk called *colostrum*, which is rich in antibodies and gives the kittens an immediate advantage in the coming struggle to avoid the diseases of infancy. It is also rich in proteins and minerals and its production lasts for several days, before the mother cat starts to produce the normal milk supply.

How do kittens avoid squabbling when feeding from the mother?

Within a few days of birth each kitten has developed an attachment to its own personal nipple, which it recognizes with ease. Amazing as it may sound, this is possible because each nipple has a special smell. We know this because if the belly region of the mother cat is washed by her human owner, so that it is cleansed of its natural fragrance, the kittens fail to find their favourite nipples. Instead of peacefully taking up their usual stations, they become disorientated. Confusion reigns and squabbling occurs.

It is remarkable to think that in the 'simple' world of the very young kitten there is odour detection based on differences so subtle that they can label each nipple as clearly as name-cards on school lockers, and that in this way orderly sharing can be maintained at feeding time.

Will one female cat feed another one's kittens?

Yes, she will. If a nursing mother has a normal-sized litter it is possible to add one or two orphaned kittens to it without much difficulty. They would probably be accepted simply by placing them, mewing plaintively, near her nest-box. Her maternal instincts would be so strong that she would be unable to resist their calls for help and would soon approach them, pick them up individually in her mouth and place them in her bed. There she would lick them and give them her scent and then allow them to feed alongside her own kittens.

Some breeders fear that this method might not always be successful, so they give the process a little assistance. They do this by waiting for the mother to leave the nest. While she is away feeding they take the strange kittens and rub them gently in the bedding that carries the female's scent. Then they leave them there, in among her own kittens, and when she returns the chances are that she will calmly lie down and let all the kittens feed from her without examining them in detail. Mother cats do not seem to be very efficient at counting their kittens and if the newcomers have become covered in the 'home scent', all is well.

Where a large number of female cats are kept together in a cattery, observers have noticed that the kittens born there are often shared out between the mothers. These group-living females show remarkable degrees of social tolerance and sometimes take up residence in large, communal nests, carrying all their kittens in there and piling them up in a huge, squirming mass. On one occasion no fewer than six females with eighteen kittens established a communal nest of this type and each female allowed the other mothers to offer their milk to any kittens whenever they felt like it. Normally, when there is a single mother cat, each kitten is the 'owner' of its own personal nipple and always returns to the same nipple every time it feeds. But in these nursery nests, the kittens took the first nipple they came across, regardless of whether it was in a familiar position on the female's belly, or even which belly it was. This free-and-easy arrangement produced strong, healthy kittens that flourished because of the division of labour of the mother cats. There was only one drawback and that concerned the weaker kittens that sometimes found themselves at the bottom of a pile of bodies and unable to breathe. An occasional casualty from suffocation was recorded, but in other respects the group maternity-home worked extremely efficiently.

In the wild such behaviour would normally never occur because of the large size of each adult cat's territory. For one litter of kittens to encounter another, or for one nursing mother to come close to another's nest, would be a rare occurrence. As a result there would have been little or no evolutionary pressure on cats to develop an anti-stranger reaction where kittens are concerned. Hence the easy sharing of kittens in a cattery where the adult females have already, through overcrowding, come to tolerate one another's presence.

Although this altruistic caring for other cats' kittens is abnormal for many felines, it does show how, under freak conditions of extreme crowding, it might be possible for a group of wild cats to start behaving like a pride of lions. Indeed, it has been suggested that this is precisely how lion prides arose, many years ago on the prey-rich plains of Africa, where the surplus of food led to an unusual increase in the lion population.

Can a litter of kittens have more than one father?

Yes, this is possible and it even has a name: *superfecundation*. A glance at the mating behaviour of cats tells why. When the female comes into heat, her calling and her sexual fragrance attract tom-cats from all around. They gather near her and squabble among themselves with much caterwauling. Then one of them approaches her and mates. The act of copulation usually only takes about five seconds, ejaculation occurring as soon as the male has entered the female. After a rest of about twenty minutes, they copulate again and this process is repeated approximately seven times, by which stage the male is usually satiated.

Some females develop a special attachment to one particular tom-cat and reject other suitors, waiting for the favoured male to become sexually aroused again. But it is just as likely that she will allow one male after another to mount her until her whole circle of admirers has been accommodated. This means that her reproductive tract will contain a mixture of sperm from several sources and it becomes almost a matter of chance as to which particular male's sperm fertilizes each of her shed eggs.

The result of this is sometimes a multi-patterned litter of kittens, which some owners mistakenly consider to be the outcome of 'genetic variety' within the make-up of their female and an unknown 'husband'. But the wildly differing kittens may instead be the product of the sexual promiscuity of their female.

This is essentially a phenomenon of domestic cats, because the territories of wild cats are so much bigger, and the chances of a whole group of tom-cats coming together in one spot when a wild female is on heat are more remote. Superfecundation is most likely to occur in town and city cats, where the individual territories have become so reduced in size that the odour of a sexually active female can easily be detected by a whole collection of different males. Were it not for the extreme aggression that subsequently erupts when the males come too close to one another, superfecundation would undoubtedly be even more common than it is, but some males will never dare to risk a mating with a dominant tom-cat watching from nearby. On the other hand, if female cats were not so sexually voracious, superfecundation would also be far less common. If the queen cat was satisfied from her seven matings with one tom, she would leave the other males in the lurch and make for home. But typically she refuses to do this, writhing on the ground and inviting more and more matings until her period of heat has passed. By this time, with the top tom-cats so sexually exhausted, even some of the masculine runts may risk a quick mating.

There is a feline phenomenon even stranger than superfecundation and that is *superfetation*. Female cats are such powerful breeding machines that some of them may even come into heat while they are pregnant. It is one of the basic rules of reproductive behaviour that the condition of pregnancy suppresses a female's sexual physiology, but female cats break this golden rule. In about one out of ten, when there is a low level of pregnancy hormone in the system, there is another phase of sexual receptivity actually *during* the pregnancy cycle. Feline pregnancy lasts nine weeks, and the additional heats usually occur after three or six weeks. These send the expectant mothers out on the tiles again where, if they are mated once more, they will be fertilized again and then carry two litters at two different stages of development.

In these cases, both sets of kittens continue to develop alongside one another, with the later group three or six weeks behind the early one. This creates two alternative problems for the mother-to-be. When she starts to give birth to the older litter, the upheaval of delivery may lead to the younger litter being ejected as well. If this happens they are so premature that they nearly always die. If, on the other hand, they manage to hang on inside the uterus, they may be born successfully at full term three or six weeks later. This causes a second type of problem – an almost impossible demand for nipples and milk supply. But if the female is able to cope with all or part of this added maternal burden she can, of course, contribute even more spectacularly to the feline population explosion.

Why do tom-cats kill kittens?

As a father, the tom-cat has a bad reputation. For centuries he has been looked upon as a sex maniac whose only interest in kittens is to kill them if he gets half a chance. This image of the male owes its origin to the writings of the great historian Herodotus, following his visit to ancient Egypt two and a half thousand years ago. Amazed by the Egyptians' devotion to their cat population, he felt inspired to comment on certain aspects of the behaviour of the felines.

Among his observations is the following assessment of the sexual cunning of the tom-cat: 'As the females when they have kittened no longer seek the company of the males, these last, to obtain once more their companionship, practise a curious artifice. They seize the kittens, carry them off, and kill them; but do not eat them afterwards. Upon this the females, being deprived of their young and longing to supply their place, seek the males once more, since they are particularly fond of their offspring.'

In other words, the sex-mad tom-cats destroy the litters of kittens in order to get the females back on heat again more quickly. This story has lasted well during the past two millennia and many people still believe it, so that tom-cats are always kept carefully away from nursing mother cats and their young kittens, in case the urge to commit lust-inspired infanticide overcomes them. Nobody has commented on any possible biological advantage of such a reaction on the part of tom-cats, or why the males should want to eliminate their own genetic progeny. So what is the truth?

Observations of European wild cats, which belong to the same species as the domestic cat, reveal that, far from being kitten-killers, the males sometimes actively participate in rearing the young. One tom was seen to carry his own food to the entrance of the den in which his female had given birth and place it there for her. Another tom did the same thing, supplying his female with food while she was unable to leave the nest during the first days after producing her litter. He also became very defensive and threatened human visitors in a way that he had not done before the young were born. Both these cat families were in zoos, where the proximity of the male was forced on the female and where, if anywhere, one might have expected to see tom-cat aggression towards the young.

In the wild, where cats have huge territories, the chances of a tom-cat coming across a female in her den with her kittens is remote, so there is little opportunity for either paternal care or paternal infanticide. In the crowded conditions of the zoo or the human city, greater proximity increases the likelihood of tom-cat/kitten encounters and when these happen one of four reactions occurs:

1 The male simply ignores the kittens.
2 The male behaves paternally towards them, as in the case of the zoo cats.
3 The female attacks the male as soon as he approaches her nest, and drives him away before he can reveal how he would have responded to the kittens.
4 The male kills the kittens.

Although this fourth reaction is the traditionally accepted one, it is in reality extremely rare. Nearly all the encounters end in one of the other three ways. But clearly the old tale from Herodotus would not have survived 2,500 years without any supporting evidence whatsoever, so how can the rare cases that have kept the story going be explained?

The answer seems to be that a female cat sometimes experiences a 'false heat' a few weeks after she has given birth. If a tom-cat is nearby this excites him tremendously, but the female usually fights with him and drives him off. Now in a great state of sexual arousal, the frustrated tom is desperate. If he meets a small kitten at this stage he may try to mount it and mate with it. The low, crouched posture of the kitten is similar to that of the sexually responsive adult female cat. This, and the kitten's inability to move away quickly when the male mounts it, act as sexual signals to the over-excited tom-cat and seal the fate of the unfortunate kitten. The male does not attack it but, when mounting the tiny animal, simply performs the perfectly normal neck-bite that he employs when copulating with a female. To the kitten this feels just like its mother's maternal grabbing, so it does not struggle. Indeed, it responds by remaining very still. This is the specific sexual signal from the adult female that tells the male that she is ready to mate. The misunderstanding causes disaster when the mounted tom-cat discovers that the kitten is too small for mating. He cannot manoeuvre himself into the correct position. His response to this problem is to grip the kitten's neck tighter and tighter, as if he is dealing with an awkward adult mate. In the process he accidentally crunches the kitten's tiny, delicate head and it dies.

Once it has been killed, the kitten may trigger off a new reaction. Dead kittens are often devoured by their parents seemingly as a way of keeping the nest clean. So the victim of the tom-cat's sexual frustration may now be eaten, as the final act of this gruesome misfiring of the feline reproductive sequence. It is these rare occurrences that have led to stories of tom-cat cannibalism – and to stories that paint the male feline as a savage monster bent on slaughtering and consuming his own children. For so many animal fallacies, it is the rare event that becomes established as the 'norm' in popular animal lore, and usually – as in this case – with the animal motives involved luridly exaggerated or distorted.

Why do white cats make bad mothers?

This is because they often do not hear their kittens calling to them and they ignore their cries for attention. However, this is not due to white cats being stupid or careless mothers, but to the fact that a large proportion of them are deaf and are therefore unaware of the problems of their mewing offspring.

Not all white cats are deaf, however, and owners of these animals should carry out some simple noise tests to find out whether they are lucky or not. It is the blue-eyed white cats that are most prone to deafness. Those with orange eyes are much more likely to be able to hear.

It is important when testing a cat's hearing to ensure that you make the noise to check its reactions out of sight of the cat. And it is also important not to make the noise by stamping or banging against a hard surface, because this can set up vibrations that even a totally deaf cat can detect through the sensitive pads of its feet. But if the cat can see nothing and feel nothing and yet it still reacts to the loud noise you make, then yours is one of the fortunate, hearing white cats.

If, on the other hand, your cat is one of the deaf ones, then there is nothing you can do to help it. Its cochlea, that vital, snail-shaped organ in the inner ear, will have started to degenerate a few days after birth, and the deterioration is completely irreversible. It is a genetically-linked defect and will be passed on to the white offspring of the deaf mother. It is therefore important not to breed from such cats if possible. In this way, the small proportion of white cats that *can* hear will become more common and the defect could, in theory, be wiped out after a few generations.

The particular combination of white fur and blue eyes seems to be the crucial one, and this is brought home vividly in the case of odd-eyed cats. Sometimes a white cat is born with one blue eye and one orange eye. In such cases, tests show that only the ear on the side of the blue eye is deaf. The ear on the side of the orange eye works perfectly well in most cases. Such cats may be at a disadvantage when hunting, because their sense of directional sound will be poor, but in other respects they can lead normal lives and make good mothers.

Owners of deaf white cats report that their pets are brilliant at compensating for their genetic disability. They become extra sensitive to tiny vibrations made by sounds and can almost 'hear through their feet'. Their watchfulness is also dramatically increased, so that they can make maximum use of their excellent sense of vision. Indeed it is not such a great tragedy to discover that a pet cat is deaf, although there is a sad weakening of the intimate communication system that cat owners and their cats enjoy sharing. But, like the cat itself, the owners can learn to become more visual in their feline exchanges, using gestures and movements where otherwise they might have used the human voice.

At what rate do kittens develop?

When they are born the kittens are blind and deaf, but have a strong sense of smell. They are also sensitive to touch and soon start rooting for the mother's nipples. At this stage they weigh between two and four ounces, the average birth weight being roughly three-and-a-half ounces. They are about five inches long.

By day four, the kittens have already started the paw-treading action which helps to stimulate the mother's milk-flow. At the end of the first week of life their eyes begin to open and they have by now doubled their body weight. As they approach the end of their first month of life, they show the first signs of playing with one another. They can move themselves about with more efficiency and can sit up. Whatever colour their eyes will be later in life, at this stage all kittens are blue-eyed and remain so until they are about three months old. Their teeth are beginning to break through at the age of one month.

At roughly thirty-two days, they eat their first solid food, but they will not be weaned until they are two months old. (Wild cats take longer to wean their kittens – about four months.) During their second month of life they become very lively and intensely playful with one another. Inside the house, pet kittens will use their mother's dirt tray by the time they are one-and-a-half months old. Play-fighting and play-hunting become dominant features at the end of the second month.

In their third month of life they are in for a shock. The mother refuses to allow them access to her nipples. They must now make do entirely with solids and with liquids lapped from a dish. Before long their mother will be coming into oestrus again and concentrating on tom-cats once more.

In their fifth month the young cats begin to scent-mark their home range. They are shedding their milk teeth and exploring their exciting new world in a less playful manner. The chances are that their mother is already pregnant again by now, unless her human owners have kept her indoors against her will.

At six months, the young cats are fully independent, capable of hunting prey and fending for themselves.

Why does a cat move its kittens to a new nest?

When the kittens are between twenty and thirty days old, their mother usually moves them to a new nest site. Each kitten is picked up firmly by the scruff of the neck and, with mother's head held as high as possible, is carried off to the fresh location. If it has to be transported over a long distance the mother may grow tired of the weight and let her head sag, switching from carrying to dragging. The kitten never objects, lying limp and still in its mother's jaws, with its tail curled up between its bent hind legs. This posture makes the kitten's body as short as possible and reduces the danger of bumps as it is unceremoniously shunted from old nest to new one.

As soon as the mother arrives at the new site she has chosen, she opens her jaws and the kitten drops to the ground. She then returns for the next kitten and the next, until the whole litter has been transported. After the last one has been moved, she makes a final trip to inspect the old nest, making doubly sure that nobody has been left behind. This suggests that counting kittens is not one of the cat's strong points.

It is usually stated that this removal operation is caused either by the old nest becoming fouled or because the kittens have outgrown it. These explanations seem logical enough, but they are not the true reason. A cat with a large, clean nest is just as likely to set about moving its litter. The real answer lies with the wild ancestors of the domestic cat. In the natural environment, away from canned cat food and dishes of milk, the mother cat must start bringing prey back to the nest, to arouse the carnivorous responses of her offspring. When the kittens are between thirty and forty days old they will have to begin eating solids, and it is this change in their behaviour that is behind the removal operation. The first, old nest had to be chosen for maximum snugness and security. The kittens were so helpless then and needed protection above all else. But during the second month of their lives, after their teeth have broken through, they need to learn how to bite and chew the prey animals brought by the mother. So a second nest is needed to facilitate this. The primary consideration now is proximity to the best food supply, reducing the mother's task of repeatedly bringing food to her young.

This removal operation still occurs in domestic cats – if they are given half the chance – despite the fact that the feeding problem has been eliminated by the regular refilling of food dishes by their human owners. It is an ancient pattern of maternal feline behaviour which, like hunting itself, refuses to die away simply because of the soft lifestyle of domestication.

In addition to this 'food-source removal pattern', there are, of course, many examples of a cat quickly transporting her litter away from what she considers to be a dangerous nest site. If human curiosity becomes too strong and prying eyes and groping hands cannot keep away from the 'secret' nest, strange human smells may make it an unattractive abode. The mother cat may then search for a new home, simply to get more privacy. Moves of this kind can take place at any stage of the maternal cycle. In wild species of cats, interference with the young at the nest may result in a more drastic measure, the mother refusing to recognize them as offspring any more, and abandoning them or even eating them. What happens, in effect, is that the alien smells on the kitten's body make it into an alien 'species' – in other words, into a prey species – and the obvious response to such an object is to eat it. Domestic cats rarely respond in this way, because they have become so used to the scents and odours of their human owners that they do not class them as alien. Kittens handled by humans therefore usually remain 'in the family', even if they have acquired new scents.

How do kittens learn to kill?

The short answer is that they do not need to learn how to perform the killing action, but it does help if they get some instruction from their mother. Kittens reared by scientists, in isolation from the mother cat, were able to kill prey when given live rodents for the first time. Not all these kittens succeeded, however. Out of twenty tested, only nine killed and only three of those actually ate their kills. Kittens reared in a rodent-killing environment, where they could witness kills but never saw the prey eaten, were much more successful. Eighteen out of twenty-one such kittens tested were killers and nine of these actually ate their kills.

Interestingly, of eighteen kittens reared in the company of rodents, only three became rodent-killers later on. The other fifteen could not be trained to kill later by seeing other cats killing. For them the rodents had become 'family' and were no longer 'prey'. Even the three killers would not attack rodents of the same species as the one with which they were reared. Although it is clear that there is an inborn killing pattern with kittens, this pattern can be damaged by unnatural rearing conditions.

Conversely, really efficient killers have to experience a kittenhood which exposes them to as much hunting and killing as possible. The very best hunters are those which, as youngsters, were able to accompany their mother on the prowl and watch her dealing with prey. Also, at a more tender age, they benefited from her bringing prey to the nest to show them. If the mother does not bring prey to the kittens in the nest between the sixth and twentieth week of their lives, they will be far less efficient as hunters in later life.

Why does a kitten sometimes throw a toy into the air when playing?

The scene is familiar enough. A kitten tires of stalking and chasing a ball. It suddenly and without warning flips one of its paws under the ball, flinging it up into the air and backwards over its head. As the ball flies through the air, the kitten swings round and follows it, pouncing on it and 'killing' it yet again. As a slight variation, faced with a larger ball, it will perform the backward flip using both front feet at the same time.

The usual interpretation of this playful behaviour is that the kitten is being inventive and cunningly intelligent. Because its toy will not fly up into the air like a living bird, the kitten 'puts life into it' by flinging the ball over its shoulder, so that it can then enjoy pursuing the more excitingly 'lively' prey-substitute. This credits the kitten with a remarkable capacity for creative play – for inventing a bird in flight. In support of this idea is the fact that no adult cat hunting a real bird would use the 'flip-up' action of the front paws. This action, it is argued, is the truly inventive movement, reflecting the kitten's advanced intelligence.

Unfortunately this interpretation is wrong. It is based on an ignorance of the instinctive hunting actions of the cat. In the wild state, cats have three different patterns of attack, depending on whether they are hunting mice, birds or fish. With mice, they stalk, pounce, trap with the front feet and then bite. With birds they stalk, pounce and then, if the bird flies up into the air, they leap up after it, swiping at it with both front feet at once. If they are quick enough and trap the bird's body in the pincer movement of their front legs, they pull it *down* to the ground for the killing-bite. Less familiar is the way in which cats hunt for fish. They do this by lying in wait at the water's edge and then, when an unwary fish swims near, they dip a paw swiftly into the water and slide it rapidly under the fish's body, flipping the fish up out of the water. The direction of the flip is back and over the cat's shoulders, and it flings the fish clear of the water. As the startled fish lands on the grass behind the cat, the hunter swings round and pounces. If the fish is too large to be flipped with the claws of just one front foot, then the cat may risk plunging both front feet into the water at once, grabbing the fish from underneath with its extended claws and then flinging the prey bodily backwards over its head.

It is these instinctive fishing actions that the kittens are performing with their 'flip-up' of the toy ball, not some new action they have learned or invented. The reason why this has been overlooked in the past is because few people have watched cats fishing successfully in the wild, whereas many people have seen their pets leaping up at birds on the garden lawn.

A Dutch research project was able to reveal that the scooping up of fish from the water, using the 'flip-up' action, matures surprisingly early and without the benefit of maternal instruction. Kittens allowed to hunt fish regularly from their fifth week of life onwards, but in the absence of their mother, became successful anglers by the age of seven weeks. So the playful kitten throwing a ball over its shoulder is really doing no more than it would do for real, if it were growing up in the wild, near a pond or river.

Why do some cats become wool-suckers?

Some cat owners notice that their cats become wool-suckers at a certain age and they worry about this apparently abnormal behaviour. They are right to do so, because it can lead to serious problems if it persists. Strands of wool are loosened by the prolonged sucking and may be swallowed, leading eventually to a blockage of the cat's intestines, requiring surgery. Why do they do it?

The clue is in the way they perform the sucking action. They find some woollen garment or other soft furnishing in the house and then settle down on it in a contented fashion. Pressing the mouth to it, they start to suck or chew it rhythmically while treading alternately with their front feet. While doing this they appear to be lost in pleasure and increasingly unaware of the world around them. Clearly there is a huge reward for them in performing these seemingly useless actions and, were it not for the risk of ingesting the wool fragments, there would be no harm in it.

The reason for the attraction of the wool is not hard to find. The actions directed towards the cloth are identical with those performed by a tiny kitten when feeding at its mother's nipple. The trampling movements are intended to stimulate her milk flow and when these are done by the wool-sucking cat they reveal that it is treating the piece of woollen material as a surrogate mother. In other words wool-sucking is the act of feeding at a 'ghost-nipple' and is the feline equivalent of thumb-sucking in human babies (or pipe-sucking in elderly human males).

Wool-sucking seems to be most common among young cats that have been orphaned or have for some other reason been deprived of the maternal nipple too soon. It usually starts shortly after the animals have been weaned and in most instances it only persists for a few months. But for some cats – especially Siamese – it continues for a lifetime and is extremely difficult to eradicate.

The special attraction of wool is that the presence of lanolin acts as a powerful unconscious reminder of the mother's belly. Sucking the wool and making it damp enhances the lanolin odour, and this keeps the cats contented and fully absorbed in their sucking and chewing.

If there is no wool available to them, cats with an urge to re-create the pleasures of sucking at the maternal nipple have been known to suck their own fur, sometimes their feet and sometimes the tips of their tails; or they occasionally develop a fixation on their owner's hair and make repeated attempts to suck on that, if they are given half a chance.

Some authorities have suggested a change in diet for such cats, but there seems little logic in this unless the new diet helps to reduce boredom. An empty lifestyle does seem to increase the chances of wool-sucking, and probably the best cure for it would be to make the cat's way of life more surprising and complex. More drastic cures, such as providing woollen garments covered in some noxious chemical substance, do not really cure the cat's fixation. All that happens is that the animal waits until the treated garments are discarded and then homes in on some other, more suitable surface. The wool-sucking then starts up once again, unabated, and it is impossible to keep up a permanent 'treatment' of all potential woollen materials. Sooner or later the cat will win and wear down its owner's resolve. The only true solution is somehow to alter the cat's mental state and rid it of the monotony or stress that drives it to perform the 'pseudo-infantile' actions.

How do cats behave when they become elderly?

Many owners fail to notice that their cats have reached 'old age'. This is due to the fact that senility has little effect on the feline appetite. Because they continue to eat greedily and with their usual vigour, it is thought that they are still 'young cats'. But there are certain tell-tale signs of ageing. Leaping and grooming are the first actions to suffer, and for the same reason. Old age makes the cat's joints stiffer and this leads to slower movements. Leaping up on to a chair or a table, or outside up on to a wall, becomes increasingly difficult. Very old cats actually need to be lifted up on to a favourite chair. As the supple quality of the young cat's flexible body is lost, it also becomes increasingly awkward for the cat to twist its neck round to groom the more inaccessible parts of its coat. These areas of its fur start to look dishevelled and at this stage a little gentle grooming by the animal's owner is a great help, even if the cat in question is not one that has generally been fussed over with brush and comb in its younger days.

As the elderly cat's body becomes more rigid, so do its habits. Its daily routine becomes increasingly fixed and novelties cause distress now, where once they may have aroused acute interest. The idea of buying a young kitten to cheer up an old cat simply does not work. It upsets the elderly animal's daily rhythm. Moving house is even more traumatic. The kindest way to treat an elderly cat is therefore to keep as much as possible to the well-established pattern of the day, but with a little physical help where required.

The outdoor life of an elderly cat is fraught with dangers. It has reached a point where disputes with younger rivals are nearly always going to end in defeat for the old animal, so a close eye must be kept on any possible persecution.

Luckily these changes do not occur until late in the lives of most cats. Human beings suffer from 'old age' for roughly the last third of their lives, but with cats it is usually only the last tenth. So their declining years are mercifully brief. The average lifespan is reckoned to be about ten years. Some authorities put it a little higher, at about twelve years, but it is impossible to be accurate because conditions of cat-keeping vary so much. The best rough guide is to say that a domestic cat should live between nine and fifteen years and should only suffer from senile decline for about the final year or so of that span.

There have been many arguments about the record longevity of a domestic cat, with some amazing claims being made, some as high as forty-three years. The longest accepted lifespan at present, however, is thirty-six years for a tabby called 'Puss' which lived from 1903 to 1939. This is exceptional and extremely rare. Serious attempts to locate cats over twenty years old, both in Great Britain and the United States, have never managed to turn up more than a mere handful of reliable cases.

One of the reasons why it is difficult to find good records of long-lived cats is that the most carefully kept details are always for the pure-bred animals, which live much shorter lives than the crossbred 'moggies' or mongrels. This is because the prized and carefully recorded pure-bred cats suffer from inbreeding which shortens their lives. The 'badly bred' alley cat, by comparison, enjoys what is called 'hybrid vigour' – the improved physical stamina produced by outbreeding. Unfortunately, such cats are less well looked after in most cases, so they, in their turn, suffer more from fighting, neglect and irregular diet. This cuts down their lifespan. The cat with a record-breaking lifespan is therefore most likely to be the one which has a dubious pedigree but is a much loved and protected pet. For such an animal fifteen to twenty years is not too hopeless a target.

One of the strangest features of cat longevity is that it easily exceeds that of dogs. The record for a dog is twenty-nine years, seven years short of the longest-lived cat. Bearing in mind the fact that larger animals usually live longer than smaller ones, the figures should be reversed, so for their size cats do unusually well. And there is a compensation for those toms that suffer the mutilation of being neutered, for neutered toms have a longer lifespan than 'entire' ones. The reasons for this, it appears, are that they get involved in fewer damaging brawls with rivals, and also that they are, for some reason, more resistant to infection. One careful study revealed that a neutered tom could expect on average three more years of feline survival than an unneutered one.

Why do cats prefer to die alone?

Cat owners are sometimes distressed by the fact that their favourite and much-loved cat leaves them shortly before it dies. After years of being cared for and protected in a human family, the elderly feline disappears one day and is then found dead in a corner of the garden shed next door, or in some even more secretive place. The owners feel spurned, wondering why their cat has not come to them for help when it feels seriously ill. To abandon them at such a moment implies that they did not, after all, mean so much to the animal — they were not a 'safe haven' in quite the way they had pictured themselves. But they do themselves an injustice.

This 'dying alone' is not a new phenomenon. An oriental author, writing as long ago as 1708, records that one of the cat's unique features is that 'it perishes in a place quite out of human sight, as if it wills not to let man see its dying look, which is unusually ugly'. Much later, a mere half century ago, the author Alan Devoe makes a similar comment: 'One day, often with no forewarning whatever, he is gone from the house and never returns. He has felt the presaging shadow of death, and he goes to meet it in the old unchanging way of the wild — alone. A cat does not want to die with the smell of humanity in his nostrils and the noise of humanity in his delicate peaked ears. Unless death strikes very quickly and suddenly, he creeps away to where it is proper that a proud wild beast should die — not on one of man's rags or cushions, but in a lonely quiet place, with his muzzle pressed against the cold earth.'

The motives described by these authors are little more than romantic inventions, but the fact that the dying cat's actions have been recorded in this way by very different writers is of some interest. Clearly we are dealing here with a feline phenomenon that is not isolated and accidental, but more a regular, typical feature of cat behaviour. If only a few cases were known, they could easily be the result of an animal just happening to die when in a remote spot. A human being who suffers a fatal heart attack when out walking in a wood would not have been setting out to 'die alone'. But with cats it seems to be too common to be explained in this way.

To understand the behaviour it is essential to consider the question of how a cat faces death. We humans all know we are going to die one day and we act accordingly. A cat has no concept of its own death and so it cannot anticipate it, no matter how ill it feels. What falling ill means to a cat, or any other animal, is that something unpleasant is threatening it. If it feels pain, it considers itself to be under attack. It is difficult for it to distinguish between one sort of pain and another, when trying to work out what is going wrong. If the pain becomes acute, the cat knows that it is in great danger. If it cannot see the source of the danger, it cannot turn to face it and lash out in defence — there is nothing there at which to lash out. This leaves only two alternative strategies: to flee or to hide. If the pain comes on when the cat is out patrolling its territory, its natural reaction will be to attempt to hide from the 'attacker'. If the cat sees a shed nearby, or some other hiding place, it will make for this and then stay concealed, alone, waiting for the threat to pass, for the pain to ease. It dare not come out, in case the source of the pain is waiting for it, and so it remains there, dying alone and in private. Despite the earlier authors' comments on the matter, at the moment of death the cat is not thinking of its human owners' feelings, but simply about how it can protect itself from the terrifying, unseen danger that is causing it so much pain.

If we feel sorry for the dying cat that cannot understand what is happening to it, we should remember that it has one enormous advantage over us: it has no fear of death, which is something we humans must all carry with us throughout our long lives.

Do cats have ESP?

Many people believe that cats are capable of some kind of extra-sensory perception, but they are not. It has become popular to explain anything puzzling as being due to some supernatural power, but this is a short cut to complacency. The scientific truth is often much more fascinating, but investigating it is stifled if we simply relegate every unusual occurrence to the dustbin of 'mystic forces'.

To start with, ESP is a contradiction in terms. Anything we perceive is, by definition, something which operates through one of our sense organs. So, if something is extra-sensory it cannot be perceived. Therefore there cannot be any such thing as ESP.

If a cat performs some very strange action – finds its way home over a long distance, predicts an earthquake, or senses the return of its owners when they are some distance from the house – then it is a challenge for us to try and find out which particular sensory pathway was involved. To put the feat down to ESP is boring, because it simply stops any further enquiry. It says the cat has magical powers, and that is an end to it. Much more stimulating is the idea that everything cats (or humans for that matter) do is capable of logical explanation, if only we can find out how the behaviour mechanism operates.

If we find that magnets attached to cats will upset their ability to find their way home, then we are beginning, very dimly, to understand the amazing homing abilities that the animals have evolved over a long period of time. If we find that cats are sensitive to very small vibrations or changes in the static electricity of their environment, then we might come to understand how they can predict earthquakes. And if we can learn more about their sensitivity to ultrasonic sounds, we may finally grasp how they can 'know' that someone is approaching from a great distance.

This does not mean that we will be able to explain everything a cat does. We may have to wait for much more advanced technology before we can do that. But ultimately it should be possible and the 'mysterious cat' will be mysterious no more. And if and when this happens, the cat will, because of our detailed knowledge of its abilities, be more fascinating to us than ever before. To explain something is not to explain it away and to understand something is not to underestimate its value.

Against all this the ESP addicts will retort that there is one feat performed by some cats that requires an acceptance of telepathic communication between human owner and lost feline. This is the 'psychic trailing' phenomenon, when a cat that has been left behind by its owners follows them to their *new* home. There are cases where owners genuinely believe that their pet cat has followed them by some mystical means to a new home several thousand miles away from the original one. The cats in question have never lived near the new home and do not have any knowledge of the new locality. Yet they turn up again, mewing to their astonished owners, sometimes as long as two years after being accidentally left at the old home site. Explain this, the ESP devotees will say, and we will withdraw our case.

If it were true it would be a remarkable telepathic phenomenon, but sadly there is usually a much simpler explanation. The world is full of stray cats, many of them searching hopefully for a new home. Anyone living in a house without a cat is likely to be visited by a stray sooner or later. There are only so many different varieties of coat colour for moggies, and if the family cat was a tabby, or a black cat with a white flash on its chest, then the owners are not to be blamed for thinking that their long-lost friend has finally found its way home. If this sounds cynical, it is not difficult to set up a simple experiment. Take a cat and place it in a special room with a circle of passages and doors. Have different people behind each door at the end of each passageway and see if the cat heads towards the door hiding its owners, ignoring the other doors with strangers behind them. When rigorous experimenting of this kind is done, the telepathic results are disappointing. So these simple tests are usually avoided by the ESP enthusiasts, who would romantically prefer the cat to remain a mystic feline force among us.

How does a cat find its way home?

Over short distances each cat has an excellent visual memory, aided when close to home by familiar scents. But how does a cat manage to set off in the right direction when it is deliberately taken several miles away from its home territory?

First of all, can it really do this? Some years ago a German zoologist borrowed a number of cats from their owners, who lived in the city of Kiel. He placed them in covered boxes and drove them round and round the city, taking a complex, winding route to confuse them as much as possible. Then he drove several miles outside town to a field in which he had installed a large maze. The maze had a covered central area with twenty-four passages leading from it. Looked at from above, the passages fanned out like compass points, at intervals of fifteen degrees. The whole maze was enclosed, so that no sunlight or starlight could penetrate to give navigation clues to the cats. Then each cat in turn was placed in the maze and allowed to roam around until it chose an exit passage. In a significant number of cases, the cats selected the passage which was pointing directly towards their home.

When these findings were first reported at an international conference most of us present were highly sceptical. The tests had certainly been rigorously conducted, but the results gave the cats a homing sensitivity so amazing that we found it hard to accept. We suspected that there must be a flaw in the experimental method. The most obvious weakness was the possibility of a memory map. Perhaps the cats could make allowances and corrections for all the twists and turns the van took as it drove around the city, so that throughout the journey they kept recalculating the direction of their home base.

This doubt was removed by some other tests with cats done in the United States. There, the cats were given doped food before the trip, so that they fell into a deep, drugged sleep throughout the journey. When they arrived they were allowed to wake up fully and were then tested.

Astonishingly, they still knew the way home.

Since then many other navigation tests have been carried out with a variety of animals and it is now beyond doubt that many species, including human beings, possess an extraordinary sensitivity to the earth's magnetic field which enables them (and us) to find the way home without visual clues. The experimental technique which clinched this was one in which powerful magnets were attached to the navigators. This disrupted their homing ability.

We are still learning exactly how this homing mechanism works. It seems likely that iron particles occurring naturally in animal tissues are the vital clue, giving the homing individuals a built-in biological compass. But there is clearly a great deal more to discover.

At least we can now accept some of the incredible homing stories that have been told in the past. Previously they were considered to be wildly exaggerated anecdotes, or cases of mistaken identity, but now it seems they must be treated seriously. Cases of cats travelling several hundred miles to return from a new home to an old one, taking several weeks to make the journey, are no longer to be scoffed at.

Can cats predict earthquakes?

The short answer is yes, they can, but we are still not sure how they do it. They may be sensitive to vibrations of the earth so minute that our instruments fail to detect them. It is known that there is a gradual build-up to earthquakes, rather than one sudden, massive tremor. It may be that cats have an advance warning system.

A second possibility is that they are responsive to the dramatic increase in static electricity that apparently precedes earthquakes. In humans there is also a response to these changes, but it is rather vague and unspecific. We speak of tenseness or throbbing in the head on such occasions, but we cannot distinguish these feelings from times when we have had a stressful day at work or perhaps when we are coming down with a cold. So we cannot read the signs accurately. In all probability cats can.

A third explanation sees cats as incredibly responsive to sudden shifts in the earth's magnetic field. Shifts of this type accompany earthquakes. Perhaps all three reactions occur at once – detection of minute tremors, electrostatic activity and magnetic upheavals. One thing is certain, cats have repeatedly become intensely agitated just before major earthquakes have struck. Cat-owners recognizing their pets' fears may well owe them their lives. In many cases cats have been observed suddenly rushing about inside the house, desperate to escape. Once the doors are opened for them they flee in panic from the buildings. Some females even rush back and forth carrying their kittens to safety. Then, a few hours later, the quake strikes and levels the buildings. This has been reported time and again from the most vulnerable earthquake areas and now serious research is under way to analyse precisely which signals the cats receive.

Similar responses have been recorded when cats have predicted volcanic eruptions or severe electrical storms. Because of their exceptional sensitivity they have often been foolishly credited with supernatural powers. In medieval times this was frequently their undoing, and many cats met a horrible death by burning at the hands of superstitious Christians because they appeared to be possessed of 'unnatural knowledge'. The fact that we now know this knowledge to be wholly natural makes it no less marvellous.

Why do we say 'he let the cat out of the bag'?

The origin of this phrase, meaning 'he gave away a secret', dates back to the eighteenth century when it referred to a market-day trick. Piglets were often taken to market in a small sack, or bag, to be sold. The trickster would put a cat in a bag and pretend that it was a pig. If the buyer insisted on seeing it, he would be told that it was too lively to risk opening up the bag, as the animal might escape. If the cat struggled so much that the trickster let the cat out of the bag, his secret was exposed. A popular name for the bag itself was a 'poke', hence that other expression 'never buy a pig in a poke'.

When was the cat first domesticated?

In most cat books, the figure given for the first domestication of the cat is between 3,500 and 4,500 years ago. This is because we have fairly strong evidence from ancient Egyptian art, showing that cats had been taken under human control by that period. It is suggested that it was the great grain stores of the early Egyptian civilization that attracted the local wild cats. The stores were overrun with rats and mice and the cats came to feast on these rodent pests, thereby endearing themselves to the Egyptian people. But there is now some new evidence that means we will have to reconsider this picture of man's first close bond with the cat.

The Egyptian connection undoubtedly occurred, but it looks as though it was quite a late development. Several thousand years earlier, man and cat appear to have already developed a special relationship, and we can now state with reasonable certainty that the cat was domesticated at least eight thousand years ago.

The evidence is slender – nothing more than a feline jaw-bone – but it is convincing for a special reason. It was discovered as recently as 1983 by Alain le Brun when excavating at the Neolithic settlement of Khirokitia in southern Cyprus and has been found to date from 6000 B.C. The important point about its location is that Cyprus has no wild cats and this means that the animal must have been brought over to the island by the early human settlers. We know that they brought other domestic animals with them but it is inconceivable that they would have taken a wild cat from the mainland. A spitting, scratching, panic-stricken wild feline would have been the last kind of boat-companion they would have wanted. Only tame, domesticated animals could possibly have been part of the goods and chattels of that early band of pioneers, striking out for a new island home.

It seems almost certain, therefore, that cats had already been tamed and domesticated on the nearby mainland by 6000 B.C., and we should not be too surprised by this. Mankind had, after all, turned from hunting to farming as a way of life at least a thousand years before this date, by which time both goats and sheep were already domesticated. Crops were already being grown and were undoubtedly attracting rats and mice in large numbers. So the cat was badly needed even at this early date. Indeed, cat bones have been found associated with human settlements such as Jericho as early as nine thousand years ago, but in those cases — until now — we could not be certain about their significance. There were plenty of wild cats living in the nearby countryside and the inhabitants of Jericho may simply have trapped or hunted these wild felines and then eaten them. The bones provided no proof of taming or domestication. But this new bone is much stronger evidence. What is more, careful examination of its proportions has revealed that it belonged to precisely the same species of cat that was later domesticated in Egypt. So the cat is, after all, not one of our more recent animal companions, but one of our oldest. And, who knows, perhaps one day in the future we will find further evidence from an even earlier date, to take the domestic cat right back to the very beginning of the Neolithic period, some ten thousand years ago.

How has domestication changed the cat?

Very little. Unlike the domestic dog, the modern cat has remained close to its ancestral form. Both in anatomy and in behaviour it is still remarkably like the African wild cat from which it was gradually developed thousands of years ago in the Middle East.

Most of the changes that have occurred since then are superficial. Coat colour has been altered and hair length has been modified, but beneath the surface even the most pampered of pedigreed cats is still the same predatory pest killer that protected the food stores of ancient Egypt.

One of the few significant alterations has been a stepping up of the breeding cycles of the domesticated breeds. Modern pet cats can easily go through three reproductive cycles in a year, whereas the wild-type will only breed once in the spring. This tripling of the breeding rate accounts to a large extent for the dramatic way in which cat populations can explode in modern urban areas.

A second change is toward a slightly smaller body size than is found among the wild specimens. Whether this was a deliberate step taken by early cat keepers in ancient times to make their new-found animal partners easier to handle, or whether it was the result of a great deal of inbreeding, is hard to say, but it is nevertheless a significant feature of feline domestication.

Third, modern domestic cats are slightly more 'juvenile' than their wild ancestors. This is the result, almost certainly, of unplanned selection by centuries of cat owners. Animals that remain playful even as adults suit us better, so we favour them. They are the ones that we are most likely to breed. They have two advantages: not only are they more attractive as companions, but they are also more easy to control. Because they are juvenilized, they are more ready to look upon their human owners as pseudo-parents. This means that they will also look upon the human home as their 'nest' long after they have ceased to be kittens. And this means that they will be more likely to return home repeatedly for parental reassurance after each of their territorial forays.

Less juvenile cats would be more inclined to wander off, abandon the parental site, and seek an entirely new territory to make their own. This is what wild kittens do when they become mature. But domestic kittens must stay put and live out their adult lives as split personalities, part breeder and mouse killer, and part pseudo-kitten toward their human family. This process has gone further and further in recent years as cats have become more important as house pets than as pest destroyers. The new, man-handled cat must be prepared for a great deal of interference. Human hands will repeatedly reach out to stroke and cuddle it. Only the kitten inside the adult cat will tolerate this. So perhaps it is true to say that the single most important change in the 8,000 years of feline domestication is the creation of this infantile-adult feline.

The cat remains, nevertheless, highly adaptable and can switch to being a full-blooded wild adult with great speed. If kittens are born to a domestic cat that has turned her back on human protection, they will grow up as untame as any truly wild ones. A farm kitten that has never seen a human being during its formative weeks will become a ball of spitting fury if cornered when it is half-grown. Great patience is needed then to convert it into a friendly adult. So the cat has it both ways: it has the capacity to be a domestic kitten-cat and has retained the option of becoming a wild killer-cat if its circumstances change. No wonder it has been so successful during the past few thousand years.

What is the history of the tabby cat?

The tabby cat is the most common colour form of all domestic breeds. It has a strange history and to understand it it is necessary to turn the clock back eight thousand years, to the moment when the first cats were being domesticated in the Middle East.

The cat from which our domesticated animals originated was the African wild cat. Despite its name, it is found over a very wide range of warm countries, including not only most of Africa but also the Mediterranean islands of Majorca, Corsica, Sardinia, Sicily and Crete. It also spreads right through Arabia and the Middle East as far as India and Turkestan. Farther north it is replaced by the European wild cat, which ranges from Portugal and Britain in the west to Russia in the east. Together they make up the species called *Felis sylvestris*.

The differences between these two races of wild cat support the idea that it was the African that originally gave rise to the domesticated feline. The European is too sturdy, with a broad head and a short, bushy tail that has a blunt, rounded tip. It is extremely difficult to tame and its shyness, combined with its great ferocity when cornered, suggests that it would have been troublesome for early man to domesticate.

The African wild cat, although bigger than the modern domestic cat, has a much less stocky body than the European wild cat, with an overall shape that is much closer to that of our familiar pet cats. Its head is more delicate and its tail less bushy. Above all, it is far less retiring than the European race and often approaches human settlements in search of the rodents that are plentiful there. One Victorian explorer reported that he could catch these cats and tether them near his food stores to keep the rats down. He claimed that they soon adapted to life in captivity and became useful pest-controllers. If young kittens were caught by local people they were said to become tame very quickly. This is a stark contrast to the spitting fury of the European wild cat. It is easy to imagine how the early inhabitants of the Middle East, and in particular the ancient Egyptians, could have converted this African race into a domestic partner and there is little doubt that this is what occurred, with the more northerly European form being left out of the story altogether in the initial stages.

The coat pattern of both the European and the African races can best be described as suppressed, weak or washed-out tabby. The pattern is there but it is not impressive. This is undoubtedly what the original domestic cats looked like and wall paintings confirm that three to four thousand years ago the Egyptian cats had light or broken stripes. So how did the full-blooded tabby pattern arise?

The answer appears to be that it arrived courtesy of the Romans. The ancient Egyptians had been trying for centuries to prevent the export of their sacred felines. But Phoenician traders were notorious for their shady dealings around the Mediterranean and if there was a precious commodity for which they could find a ready market, nothing could stop them. Cats soon found their way out of Egypt, smuggled carefully to Greece and later to Rome. From there, they spread across Europe as the Roman empire began to swell. Along the way, these early pest-controlling cats started to encounter the wild European cats and to hybridize with them. The result of this injection of European blood was the original full-tabby cat. Tests have since shown that when the weak-tabby European and African wild cats are crossed with one another, the hybrid kittens develop coat patterns which are much closer to the tabby patterns of modern domestic cats than they are to the markings of either of their parents. This, it seems, is how the history of the tabby began.

The first cats of this type were what we today call striped or mackerel tabby, covered with thin, dark lines. Some of these lines break up into dashes or spots, but the overall effect is of a tigerish striping. To begin with, this was the only such pattern in existence, but then a new mutation arose. A blotched tabby appeared on the scene. On this animal the markings were much bolder and more complex. The narrow striping survived only in certain areas. It is believed that these blotched tabbies arose first in Britain, in the Elizabethan era. It was a time of great British expansion and it is thought that, in the guise of ships' cats, they were scattered from the British Isles all over the globe in a comparatively short space of time. With the growth of the British Empire in Victorian times, they spread still farther.

For some reason we do not fully understand, the blotched tabby cat was a winner. Perhaps the gene for the pattern was linked to an unusual level of aggressiveness or assertiveness, with the result that these cats soon managed to oust most other colour forms whenever there was a dispute over territory or females. Perhaps they were simply more healthy or more fertile. Whatever the reason, this new pattern began to dominate. The earlier striped tabby went into a rapid decline. Today it has become quite rare, while the blotched tabby is the most common colour form of all.

It would not be too far from the truth, as one author put it, to christen this most successful of all cats as 'The British Imperial Cat'.

How did the Manx cat lose its tail?

The tailless Manx cat is one of the oldest breeds known, having appeared on the Isle of Man at least four hundred years ago. In legend, it lost its tail because it was the last animal to board the ark. With the great flood rising fast, Noah slammed the door of the ark shut in such a hurry that he accidentally severed the cat's fine bushy tail.

In reality, the tailless condition is a genetic deformity of a rather serious kind. The gene that causes the taillessness also distorts the rest of the spine, giving the cat a backbone with fewer and shorter vertebrae. In severe cases it gives rise to the condition known as spina bifida. It also gives the animal un-usually short front legs and long hind legs, and an abnormally small anus, so that it hops along like a rabbit and suffers from constipation.

There is also a lethal factor in the Manx gene. If two tailless Manx cats are mated, the kittens are so deformed that they nearly always die before birth. So Manx kittens are usually produced by crossing tailless cats with tailed ones. This gives rise to litters in which there are some *rumpies* (with no tails at all), some *rumpy-risers* (with a knob of a tail), some *stumpies* or *stubbies* (with very short tails) and some *longies* (with almost full-length tails).

How the Manx cat first arrived on the Isle of Man is hotly debated. One favourite story is that they were brought there from Japan by Phoenician traders thousands of years ago. This is based on the existence of a nearly tailless cat found in Japan – the Japanese Bobtail. Unfortunately for this theory, the genes carrying tail-reduction in the two cases are quite differ-ent, and the Manx and the Bobtail bear only a superficial resemblance to one another.

A second scenario pictures the original Manx cat swimming gamely ashore from one of the Spanish ships of the Armada escaping from the English fleet in 1588. The captain of this ship, we are told, ran his vessel aground on what is now known as The Spanish Rock and one or more tailless cats clambered to safety there. These cats were supposed to have been obtained by the Spanish captain from some unspecified source in the Middle East. A less glamorous version of its origin sees the first Manx cat as nothing more than a Manx sailor's pet — a curiosity brought home after travels to the Orient.

While everyone puzzles over how these animals first arrived on the Isle of Man, stories of tailless cats being found in such faraway places as Russia, Malaysia, and China all leading to new theories, nobody seems to have considered that the tailless gene might first have occurred on the Isle of Man itself. As a sudden mutation there is no reason whatever why it should not crop up anywhere, and the small island between Ireland and England is as likely a place for it as anywhere else.

Note:

Since writing these words I have been assured by Manx Cat specialists that the animal's troublesome genetic defects have now been largely eliminated by careful selective breeding without losing its uniquely tailless condition. If this is the case, then the future looks bright for this fascinating and historic feline.

Why do so many cat breeds come from the East?

If you visit a major cat show today you will find that there are about fifty breeds on display there, in addition to the many different colour forms and 'sub-breeds', all minor variations of well-established breeds, which need not concern us here.

A survey of the country of origin of each breed reveals a strange distribution. Although Europe is the original home of the cat show and of competitive, controlled breeding, remarkably few of the fifty basic breeds come from that part of the world. Most of the old breeds come instead from the East. Before asking why, here is a summary of the facts:

Europe boasts the British Shorthair and its French counterpart, the Chartreux; also the large Norwegian Forest cat and, from the Isle of Man, the famous tailless Manx cat. Very recently, in the 1950s and 1960s, there have been the Rex cats and the Scottish Fold cat, and even more recently the Somali and the Burmilla, although these are little more than crosses. And that is about it.

The East, on the other hand, has given us the long-haired Persian cat from sixteenth-century Iran, the silky Angora cat from Turkey at the same period (when Ankara was pronounced Angora), and the exotic, long-legged Siamese cat from seventeenth-century Thailand. Other Oriental breeds include the Birman and the Burmese from Burma, the Turkish Van cat, the Japanese Bobtail, the Egyptian Mau cat, the Abyssinian cat from what is now Ethiopia, the Russian Blue, the Singapura from Singapore, and the Korat from Thailand.

Ignoring the very recent breeds, it is clear that in earlier days, when cat-showing was just beginning, the major pedigree lines nearly all stemmed from the Near or Far East, with the West lagging far behind. The reason for this is that, in Europe, the long centuries of persecution by the Christian Church had left the Western cat with a very low status. At worst it was looked upon as evil and at best no more than a

173

utilitarian pest-destroyer. Only when we reach Victorian times do we find its status rising again, thanks to the sentimentality of the period. The climax of this change of mood was the holding of the first competitive cat show, at Crystal Palace in London in 1871. There were twenty-five classes and the animals were divided roughly into 'Eastern' and 'British'. As the years passed, the exotic Eastern cats gradually ousted the less exciting British cats and became the hot favourites, much to the distress of the British organizers.

The prominence of the Eastern cats was due to their more sacred and revered role in society. They had escaped the witch-hunt purges of Europe and had an entirely different significance in the lives of the Eastern peoples. Under the more favourable conditions of their past, they had developed into the remarkably beautiful and striking breeds that made such an impact on the visitors to the early cat shows. Even the very best British tabby had a difficult time competing with such dramatic animals.

If Europe had not suffered so many years of religious cruelty and persecution, the home-grown cats might have been better developed and ready to compete with the extravagant foreigners. But so few local breeds had undergone any kind of specialized breeding to improve the stock, and the Western animals had to attempt to catch up with their long-established and highly pampered Oriental counterparts.

Switching to modern times, the source of new breeds has moved to the North American continent. Cats developed in the United States and Canada include: the Peke-faced cat, the Ragdoll cat, the Balinese, the Maine Coon, the American Shorthair, the Snowshoe, the Havana Brown, the Ocicat, the Malayan, the Tonkinese, the Bombay, the Wirehair and the extraordinary Sphynx. Despite their often Eastern names, these are all Western pedigree cats, almost all of which have been specifically developed for the show-ring, rather than for any other purpose.

Africa, South America and Australia appear to have contributed little or nothing to the pedigree cat world. And Europe, with its disgraceful feline past, has provided far less than its fair share. In general, we have to thank the Orient for the early breeds and North America for the late ones.

How does temperature affect the colour of a cat's fur?

When we look at the colour pattern of a cat we automatically think of it as something the animal inherits from its parents. In our minds it is a fixed pattern that is uninfluenced by variations in the animal's personal environment. It is a case of 'born a black cat, always a black cat', or 'born a sealpoint, always a sealpoint'. So it comes as a surprise to discover that this is not always the case. In one major group of cats – the Siamese – the colouring of the animals depends on the temperature in which they live.

When a Siamese kitten is born its fur is all-white. As it grows, this plain colouring begins to change. Dark pigments appear at the tip of the nose, around the edges of the ears, at the very tip of the tail and on the pads of the feet. These darkening extremities, or 'points', then slowly spread as the young cat becomes mature and by the end of the first year the Siamese has its adult pattern, with the nose colour covering most of its face, the dark ear-fringe spreading to include most of its ear surface, the tail-tip colour extending right up to near the base of its tail, and the pigment at the feet stretching halfway up its legs. At the places where the dark zones meet the pale central part of the body there is a soft intermediate zone, rather than a sharp edge.

This coloration is extremely attractive to human eyes and most people think of it as just another feline coat pattern, like tabby or tortoiseshell, but it arises in a completely different way. This is shown by looking at what happens when a Siamese kitten is reared in a very cold environment. Such an animal, born white as usual, darkens dramatically as it grows older. Instead of having a pale body with dark points, it becomes dark all over. Another Siamese kitten, reared in an unusually hot environment, develops an adult coat that is pale all over, lacking the dark points altogether.

The explanation of these variations is that, in the Siamese

cat, a lower skin temperature causes more pigmentation to be laid down in the growing hairs. This is why the newborn kitten, hot from its mother's womb, is white all over. Then, as it grows up in a normal, average temperature, the hotter area of its body – its central trunk region – remains pale in colour, while its cooler extremities become gradually darker. There is just enough difference between, say, the tip of the tail and the soft flanks of the cat, to produce the typical two-tone pattern of the Siamese. In fact, to study the sealpoint Siamese coat pattern is to observe what amounts to a temperature map of the animal's body surface.

Interestingly, if a Siamese cat is injured on one leg or its tail, and the damaged part is bandaged up while it recovers, this will have an impact on its coat colour. The hairs that start growing under the bandage – where the encased skin is hotter than usual – will lack pigmentation. When the bandage is removed there will at first be no sign of this. The old, dark hairs do not, of course, change. But later on, when the hairs that began sprouting in the heat of the bandage become fully grown and replace the old hairs, a startling white patch will appear. For pet cats this can be strangely disfiguring, upsetting the neat balance of the Siamese sealpoint pattern, and for pedigree show cats it can be disastrous, robbing them of any chance of winning a competition, at least until another whole growth of hairs has occurred and the white ones replaced in their turn.

Sometimes a Siamese cat will develop pale colouring on its points without any local injury. When this happens it means that the animal has suffered some kind of temperature-increasing illness, such as a prolonged fever, or possibly some kind of shock or trauma. For these reasons, the Siamese has to be kept calm and healthy if it is to retain its beautiful sealpoint pattern. Even with healthy Siamese, however, there is a problem for the pedigree exhibition cat, because as an animal grows older its general body temperature begins to fall very slightly, causing its body fur to darken little by little. As a result, a champion Siamese cat's career is usually over after only three or four years, as the hot-blooded youngster matures gracefully into the cool cat of older years.

Which are the giants and midgets of the cat world?

The size of domestic cats varies far less than that of domestic dogs. Dogs have been bred for many different tasks, from the massive guard dogs and fighting dogs right down to the little toy dogs and lap-dogs. But the cat has only had one main task throughout its long history of domestication and that is vermin-killing. This occupation has had little influence on its size. The North African wild cat weighs only slightly more than the average moggie.

So the giants and the midgets of the cat world are not particularly impressive. They do exist, however, because certain breeds of cat have become adapted to different climates. As with all animals, the colder the climate the bigger the body becomes. Modern breeds derived from northern European cats should therefore be heavier than those from the tropics. This is indeed what we find. The British Shorthair, descended from generations of felines that struggled to survive in the bleak climate of their island home, is stocky and sturdy and much more massive in build than, say, the lean-bodied Siamese from the steamy heat of Bangkok. Fully-grown tom-cats of these two contrasting breeds would weigh approximately 12 lbs and 9 lbs respectively.

A typical moggie is intermediate between these two extremes, with a tom weighing in at about 10 lbs and a queen at 8 lbs. A few freaks have been discovered from time to time, one amazing animal tipping the scales at no less than 43 lbs and a dwarf specimen at as little as 3 lbs, but these were abnormal. The giant was suffering from hormonal imbalance and the midget was a case of genetic dwarfism.

Among the pedigree breeds, the Maine Coon cat from North America is one of the biggest, with some exceptional individuals weighing as much as 30 lbs. The Norwegian Forest cat is another large breed, coming as it does from the coldest part of the domestic cat's range.

The smallest breed is one that has only recently arrived from Singapore, where it was known as the 'drain cat'. Cats are not particularly popular there, it seems, and this particular breed became smaller and smaller as an adaptation not only to the hot climates but also to the need to find restricted hiding-places. It was not officially 'discovered' until the 1970s, when a few were brought to the United States. There it is becoming increasingly popular, presumably because its diminutive body suits the increasing shift towards apartment-dwelling among modern Americans. Male Singapuras weigh, on average, only 6 lbs, and the females a mere 4 lbs.

When did cat shows begin?

The first recorded cat show took place at the St Giles Fair in Winchester, in 1598, when a few specimens were put on display. Similar small shows were staged at similar fairs, but these were of no significance and had no official status. The breeding of pedigreed cats had little meaning at this time.

It was not until the second half of the last century that serious, competitive cat showing was staged. The earliest example was at a London house in 1861, but this was still not a true public exhibition. Similar developments were taking place in the United States, where a small show featuring the Maine Coon cat, in particular, was put on in New England in the late 1860s.

Then, in 1871, the first major official cat show was staged. It was held at the Crystal Palace in London and was organized by an author and artist named Harrison Weir. It opened on Thursday, 13 July, and there were so many visitors that the cats themselves were barely visible in the dense throng. Weir was amazed by the public response. On the journey to the Crystal Palace he had serious doubts about the wisdom of his plan, fearing that the animals would 'sulk or cry for liberty'. But when he arrived he found them lying peacefully on crimson cushions. There was no noise except for widespread purring and gentle lapping of fresh milk.

A total of 170 cats were entered, although the prize money amounted to less than £10 for the whole competition. One of the prize winners was Harrison Weir's own fourteen-year-old tabby called The Old Lady, which is not so surprising when one discovers that two of the three judges were Weir's brother and Weir himself. If this system of judging left something to be desired, it must be recorded that, thanks to Harrison Weir, cats were, for the very first time, given specific standards and classes. These form the basis of the system still employed today at modern cat shows all over the world.

Further shows were soon organized, and in 1887 the

National Cat Club was formed to rule this new competitive world. The president was, needless to say, Harrison Weir. By 1893 the first official cat stud book had been started, and pedigreed cat breeding had begun in earnest. In 1895 Madison Square Garden in New York saw the first major American cat show, although this was organized by an Englishman. Today there are no fewer than sixty-five annual shows in Britain and four hundred in the United States. Unfortunately, the world of cat show organizers has become increasingly competitive within its ranks, and there have been splits and divisions since the earliest days. The result is that there is no single authority in either Great Britain or the United States, and each club or society has its own slightly different rules and classes. All of this can be confusing to the outsider, but for the cats it does not make a great deal of difference. The survival of the pedigreed competition is the important thing, maintaining the serious attitude toward pure-bred felines and preventing cats from losing status in twentieth-century society.

With more than 94 per cent of cats today being non-pedigreed, the cats of the show world constitute only a tiny minority of the general feline population, but this does not matter. As long as the élite pedigrees exist to be photographed and exhibited, they will transmit an aura of importance to domestic cats in general. They are the ambassadors of the feline world, and when over 2,000 of them gather each December for the biggest cat show in the world – the National Cat Show in London – the interest they arouse is an admirable advertisement for the value we place on our feline companions.

Are tortoiseshell cats always female?

Nearly always. The chances of finding a male tortoiseshell have been calculated at about 200 to 1. They may be extremely rare but they do exist. This poses two questions: why are almost all tortoiseshells females, and why, if males are so unfavoured in this colour-form, do they appear at all?

First, what exactly is a tortoiseshell cat? Although there are some variations, it is essentially a cat that displays an irregular pattern of black and red fur. 'Red' is the technical term for what most people would call ginger, orange or marmalade. Red cats are essentially red tabbies, with the red coloration in dark and light bands rather than uniformly spread. So a tortoiseshell cat is in reality a cat that is part black and part red tabby, giving it a tri-coloured look.

What makes the sex distribution of these cats so odd is that normally only a female kitten can display black patches inherited from one parent and red tabby patches inherited from the other. This is because the genes controlling these particular colour-forms are both carried on the X chromosomes, the red gene on one and the non-red gene on the other. The catch is that only females have two X chromosomes, so only females can display the red plus non-red tortoiseshell combination. Males have instead one X chromosome and one small Y chromosome, which means that on their single X they carry either the red *or* the non-red gene, but cannot have both. So they are either all-over red tabby or all-over black.

If this is the case, then how can any male tortoiseshells exist at all? The answer is that occasionally there is a minor genetic error and a male cat develops with the combination XXY. The double X gives it the chance to be red and black, while the Y gives it male characteristics. It does, however, have a problem because its masculinity leaves a lot to be desired. To start with, it is sterile. Also its behaviour is extremely odd. It acts like a masculinized female rather than a true male.

One particular male tortoiseshell that was observed in a

colony of cats revealed a strange personality. It was nonchalant in its dealings with other cats, disdainfully ignoring the usual status battles, which were nearly always between males or between females – there was little social fighting across the genders. Perhaps because the tortoiseshell male was neither fully male nor fully female, it did not feel the need to compete in these single-sex pecking-order disputes.

In other respects it was also peculiar. It did not start to spray urine at the age when any typical male would have done so. It did not court or attempt to mate with females on heat, even though it appeared to be anatomically well equipped to do so. It did, however, allow young tom-cats to mount and attempt to mate with it.

When it had grown older it did show a little interest in females and even deigned to mate with a few, though never with much enthusiasm. It also sprayed urine in a desultory fashion, but never behaved like a full-blooded tom at any stage. Once it was experimentally isolated with a highly-sexed female and was observed to mate several times, but the female failed to become pregnant, confirming the typical male tortoiseshell infertility.

So, although it is not true to say that *all* tortoiseshell cats are females, it is true to say that they are all feminine – even the rare males. And it is probably true to add that no tortoiseshell cat has ever fathered a litter of kittens.

There is one compensation, however, for the unfortunate tortoiseshell toms. Their great rarity has given them a special value in times past, so that they have often escaped the indifference and persecution that has befallen the common-place moggies. In Celtic countries it was always considered a good omen if one of these cats decided to settle in the home. In England there was a belief that warts could be removed simply by rubbing them with the tail of a tortoiseshell tom during the month of May. And Japanese fishermen would pay huge sums for a tortoiseshell tom, to keep as a ship's cat, for it was thought it would protect the crew from the ghosts of their ancestors and the vessel itself from storms.

So, although these cats may be doomed to a disappointing sex life, in other respects they have fared remarkably well.

Are some breeds abnormal?

There is a great deal of argument about what constitutes an abnormality in the world of pedigree cats. A new colour form creates no problems, but when a mutation occurs that alters the anatomy of the cat in some way, there is often a heated debate as to whether the new variant should be encouraged or allowed to die out. If it puts the cat at a major disadvantage then the answer is obvious, but if it is only a minor disadvantage then the cat breeders split into two warring camps. The result is that one official feline organization will recognize the new mutation as an additional breed, while another official body outlaws it and refuses to allow it to enter its cat shows.

At the present time there are four unusual breeds that fall into this category – accepted by some, rejected by others – and they are the Scottish Fold cat, the Canadian Sphynx cat, the Californian Ragdoll cat and the American Peke-faced cat. The Peke-faced cat was known back in the 1930s, but the other three were all discovered in the 1960s and were quickly established by enthusiastic local breeders, delighted to be founding new lines of pedigree cats. As there are fewer than fifty major breeds (ignoring all the colour variants), the discovery of entirely new types was extremely exciting, and the intense interest aroused by them is easy to understand. But, in the euphoria of the moment, were the local breeders blinding themselves to the fact that what they were really doing was preserving freaks? The only way to decide is to assess the situation in each case in terms of the cat's quality of life.

First, the Scottish Fold. There are many breeds of flop-eared dogs, but this is the only breed of cat that lacks the typical, pricked ears of the feline head. Its name refers to the fact that its ears are folded forwards so that, from the front, they appear to be horizontal in general shape, rather than vertical. The Fold cat was first spotted by a Scottish shepherd on a farm in 1961. He immediately realized that this was a most unusual kind of cat and started a breeding programme. The round-headed silhouette of the Scottish Fold gained it many admirers, although some felt it looked too sad. It was later discovered that some Scottish Folds developed thickening of the limbs and tail, and that this was a serious handicap to them. Such specimens have not been bred from since their problem was discovered, and the breed is now well established with no apparent weaknesses. It has become extremely popular in the United States, but certain official bodies are still banning it because, they say, the cat might suffer from ear mites or from deafness. The breeders of Scottish Folds retort that there is no evidence for this, and there the matter rests.

From the cat's point of view, the folded ears have the slight disadvantage that they do not communicate the usual mood signals seen when a cat, becoming angry or scared, starts to flatten its ears ready for fighting. It is the shift from fully erect ears to fully flattened ones that transmits the all-important social signal. The Scottish Fold cat appears as though it is permanently in the act of lowering its ears. This should make it look like a cat about to fight, but strangely it does not. The reason is that the folding of the ears brings them forward and this places them in a posture that is not part of the usual ear-lowering signal. In a normal cat, ears flattened to this degree would already be twisted round to the rear. So the Scottish Fold has a unique, 'squashed' ear, as distinct from a 'flattened' ear posture. Whether the cats themselves make this distinction no one seems to know. If they do, then there is no reason why this breed should not take its place among the other pedigree types.

Second, the Canadian Sphynx. This is a naked cat which was first discovered in Ontario in 1966. Apart from the fact that it lacks a coat of fur, it is a perfectly normal cat, with a charming disposition. But its nakedness does make it appear extremely ugly to many cat enthusiasts and does not seem to compensate for its oddity-value. The only serious objection to it is that it must suffer badly from the cold and, in its country of origin at least, it must spend the whole of the winter either indoors or in a man-made coat. If it has loving owners who can afford central heating, this breed presents no problems, but it has remained largely outlawed by the various cat societies.

Third, the Californian Ragdoll. This is a cat with a limp

body that genetically lacks the usual defensive reactions of felines. If picked up it hangs limply in your hands like a lifeless ragdoll and never tries to struggle. The fear for it is that it could easily be hurt without complaining and could suffer at the hands of careless children who would treat it too much like its namesake. What saves an ordinary cat from being over-mauled is the fact that sooner or later it struggles and scratches. But the Ragdoll, lacking these responses, could be literally loved to death by unthinking children. As a result, it too is banned by many cat societies, who would like to see it disappear altogether. The pity of it is that this happens to be as beautiful a cat as the Sphynx is ugly. Its defenders say that they will always keep its price high and carefully screen potential owners, as a way of protecting it. But how long could they keep control of the situation? This remains to be resolved.

Finally, the American Peke-faced cat is a long-haired breed that has been bred for a flatter and flatter face, rather as the Pekinese has been exaggerated among the dog breeds. As a result it suffers from problems with its eyes, its teeth and its breathing. Because of constriction beneath its eyes, its tear ducts can become blocked, causing runny eyes. Because of its reduced jaws, its teeth often fail to meet properly when the mouth is closed. And because of its reduced nasal cavities it may find breathing increasingly difficult as it grows older. Despite this, it does have an immensely attractive head-shape when viewed anthropomorphically. Its flattened face and its long, soft fur give it the rounded appearance of a 'super-baby', appealing to many humans. Because of this it has been popular at cat shows in the United States for several decades. But it still remains to be accepted in Britain as a suitable pedigree cat for championship status.

These four 'abnormal' breeds are all comparatively new, so they still have an uphill struggle to gain worldwide recognition. The long-established Manx cat is certainly as abnormal as any of them, with its strangely abbreviated backbone and the problems this causes, but its presence at cat shows was accepted so long ago that nobody now objects to its inclusion. And there would be a great sadness if it vanished, because it has become part of feline history.

Of the four new breeds, three are in no trouble providing

they are well looked after. If the Scottish Fold has its ears cleaned, the Canadian Sphynx is kept warm and the Californian Ragdoll is kept away from juvenile tormentors, they can all lead contented and fulfilled lives. The Peke-faced cat is another question. No matter how much love and attention this animal has, there is a danger that its respiration will suffer. So, for the sake of the animals themselves, this breed should be brought back slightly from its extremes of exaggeration. It will still be possible to have a charmingly flattened face, which transmits all those appealing baby-signals, even if not quite *so* flat. Moderation in all things has generally been a vain hope where human competition is concerned, and the contest to obtain the ultimate Peke-faced specimen has already led the breed into trouble. But the authorities of the cat world do seem to have their eyes firmly fixed on possible dangers, and we can hope for a much greater restraint than has existed in the case of a number of breeds of dog.

Are there ideal cats for allergy sufferers?

Up until the 1950s, it was difficult for any human beings who suffered from an allergic response to cats to become cat owners, no matter how much they longed to share their lives with a feline companion. But since then things have changed, and if they are prepared to go to a little trouble and expense they can now acquire a cat that will almost certainly give them no problems.

There are now two breeds of cat available that do not appear to cause allergic responses, which with an ordinary cat would lead to breathing difficulties and an asthmatic condition. The two breeds in question are the Cornish Rex and the Sphynx.

The Cornish Rex was discovered in 1950 when it was noticed that a cat with the unusual name of Kallibunker did not have normal feline fur. Kallibunker's coat was short, sparse and curly, and it was quickly realized that he might be the forerunner of an entirely new breed of pedigree cat. This is indeed what happened and, from his humble Cornish origins, Kallibunker became the founding father of the Rex dynasty.

The remarkable feature of Kallibunker's fur was that it completely lacked the usual long guard hairs found on all other cats. These are the hairs that appear to cause the allergic response in human sufferers, so it looked as if the Rex would become the ideal breed for such people to own. Several who tried keeping the cats reported delightedly that this was the case. Whether *all* asthmatics and allergy sufferers would find the Rex cat suitable has yet to be proved, since the breed has remained comparatively rare. This is because, despite its great charm, it does look rather odd to most cat-lovers.

Those who have been able to overlook its skimpy fur, which gives it a gangling, half-naked appearance, insist that in personality it is a sheer delight, retaining an almost kittenish playfulness even when adult. Some owners have claimed that their Cornish Rex cats wag their tails when pleased, like dogs, although this is hard to believe. Nevertheless, this supposed trait and their tight, curly hair have given rise to the use of the name 'Poodle cats' as a popular term for them.

Amazingly, cats with similar genes were found in Germany and when these were crossed with the Cornish Rex specimens they produced typical curly offspring. So the German Rex would also be a suitable cat for an allergy sufferer. This does not apply, however, to the Devon Rex cat, another curly-coated breed, discovered in 1960, not far from the original Cornish Rex home. Despite its close proximity, it was discovered that this breed owed its curly coat to a different gene and close inspection revealed that, unlike the Cornish and German breeds, this one did have a few long guard hairs. It would probably still be better for an asthmatic than other, fully-furred breeds, but there is little point in taking a risk with it when there is the more suitable Cornish Rex available.

If even the Cornish Rex causes doubts and worries, then there is only one option left: the bizarre but very friendly Sphynx Cat. This is an almost naked cat, and even the most sensitive allergy sufferers would be unable to generate much of a reaction to this remarkable animal. It originates from Canada, where a hairless kitten was born in 1966 and caused a sensation. In the resulting breed, the body is covered with a short, soft down in place of the usual fur and as the cat grows up this down persists only on the extremities. As a result, its face appears to be covered in suede or velvet and it feels like soft moss to the touch. The naked skin of its trunk reveals every wrinkle and crease as it moves about and to many eyes this makes it the ugliest cat in existence. But it is still a cat and a particularly affectionate one, making a most rewarding pet if you can overlook its strange nudity. One solution to this, of course, is to make it a coat of some non-allergic material. This would have the double value of hiding its stark skin and of giving it back some of its missing insulation. Definitely not a cat for someone whose house lacks central heating, but certainly an excellent breed for the allergic and the asthmatic would-be cat owner.

What is the cause of cat phobia?

For those of us who live with a friendly cat it is hard to understand people who suffer from a terror of encountering felines at close quarters. This fear of cats, or *ailurophobia* to give it its technical name, is rare, but when it does occur it can cause untold misery for the sufferer and it is worth examining how it begins and how it can be cured.

One way it can start is through a childhood trauma — a sudden unpleasant shock involving a cat or a kitten. When we are very young we see a fluffy kitten as something especially soft and cuddly and we have a strong desire to pick it up and hold it tight. Kittens sometimes take badly to this over-zealous embrace and strike out with their sharp claws. The idea that something so apparently soft and harmless has pain-inflicting daggers on the ends of the feet is enough to disturb certain infants and to make them distrust the approaches of all felines. There is a kind of infantile betrayal — the kitten says, 'I am all soft,' and then, when this statement has been fully trusted by the child, the animal strikes out painfully and draws blood. This is so unexpected when it is encountered for the first time that it feels like a deliberate deception. Distrust can then rise to the level of terror if cats are subsequently greatly avoided, so that they never become familiar to the growing child. Soon a full-blown phobia has developed.

A second way in which cat phobia can arise stems from an irrational fear on the part of parents that the family cat may try to smother the newly-arrived baby, by sitting on top of its face while the child lies sleeping in its cot. This old wives' tale is amazingly persistent, despite the fact that no cat could possibly relax and sleep on top of a squirming, suffocating baby. As a result, many an infant may experience a shrieking mother rushing into the nursery and yelling at the cat to leave the room. These early associations between cats and panic may leave their mark and resurface later in the life of the child.

These explanations cannot be the whole answer, however,

because it still remains to be explained why it is that human females are much more prone to ailurophobia than human males. There must be some hidden sexual factor involved, as though the soft-furred (and therefore sexy) cat has come to stand for sexual violence and rape, via its savagely sharp claws and canine teeth. This more psychoanalytical explanation may seem far-fetched, but it is important to realize that feline terms have often been given some sort of sexual connotation (sex kittens, pussy as slang for female genitals, and cathouse for brothel). If the cat is sexy, then fear of cats could have something to do with an abnormally suppressed sex drive.

How does the cat-hater show the feelings of fear? And what is it about the cat that is the specific trigger for the irrational panics? One important factor seems to be the tendency of the cat to 'jump up unexpectedly' and to behave in an unpredictable way when at close quarters. This feature of feline behaviour is mentioned time and again in investigations of cat phobics. Sadly for them, their response to it works against them and encourages the very behaviour they fear most. Because they are so terrified, they sit very, very still. But cats much prefer placid, stationary bodies on which to fall asleep. So the more frozen in horror a human being is, the more likely the cat is to leap up and try to settle down on the static lap. When this happens to a true phobic the results can be dramatic with screams and sometimes outbursts of weeping.

It has been suggested that the special fear of the cat jumping up on them in an unexpected way is the result of cat phobics' general dislike of spontaneity and fear of the suddenly surprising. But it seems more likely that this fear has more to do with the childhood horror of seeing the parent scream at the 'smothering' cat that has just jumped up on to a cot or bed.

The cure for cat phobia is straightforward enough, but distressing for the patient. It requires a series of step-by-step familiarization 'lessons', in which at first things only remotely feline are presented to the victim. These may be simply photographs of kittens or cats, or toy animals. After a while a kitten is placed in a small, secure cage and left on the far side of a room, while the phobic is gently reassured that it cannot get near. Gradually, the animal is moved nearer and day by day the phobia can be reduced in intensity until eventually the victim can actually hold a kitten. After this, the longer spent in the company of cats the better, but always with the careful avoidance of any sudden, unanticipated move. After a few months of therapy it is usual for even the most intense form of cat phobia to disappear. Sadly, many sufferers believe that there is no cure and can never be one. For them there is a needless, lifelong fear of encountering a strange cat, a fear that sometimes ends with them refusing to go out of doors at all. Their condition is beyond reason, but it is certainly not, as they believe, beyond cure.

Why does a cat tear at the fabric of your favourite chair?

The usual answer is that the animal is sharpening its claws. This is true, but not in the way most people imagine. They envisage a sharpening-up of blunted points rather in the manner of humans improving the condition of blunted knives. But what really occurs is the stripping-off of the old, worn-out claw sheaths to reveal glistening new claws beneath. It is more like the shedding of a snake's skin than the sharpening of a kitchen knife. Sometimes, when people run their hands over the place where the cat has been tearing at the furniture, they find what they think is a ripped-out claw and they then fear that their animal has accidentally caught its claw in some stubborn threads of the fabric and damaged its foot. But the 'ripped-out claw' is nothing more than the old outer layer that was ready to be discarded.

Cats do not employ these powerful 'stropping' actions with the hind feet. Instead they use their teeth to chew off the old outer casings from the hind claws.

A second important function of the stropping with the front feet is the exercising and strengthening of the retraction and protrusion apparatus of the claws, so vital in catching prey, fighting rivals and climbing.

A third function, not suspected by most people, is that of scent-marking. There are scent glands on the underside of the cat's front paws and these are rubbed vigorously against the fabric of the furniture being clawed. The rhythmic stropping, left paw, right paw, squeezes scent on to the surface of the cloth and rubs it in, depositing the cat's personal signature on the furniture. This is why it is always your favourite chair which seems to suffer most attention, because the cat is responding to your own personal fragrance and adding to it. Some people buy an expensive scratching-post from a pet shop, carefully impregnated with catnip to make it appealing, and are bitterly disappointed when the cat quickly ignores it and returns to stropping the furniture. Hanging an old sweat-shirt over the scratching-post might help to solve the problem, but if a cat has already established a particular chair or a special part of the house as its 'stropping spot', it is extremely hard to alter the habit.

In desperation, some cat-owners resort to the cruel practice of having their pets de-clawed. Apart from the physical pain this inflicts, it is also psychologically damaging to the cat and puts it at a serious disadvantage in all climbing pursuits, hunting actitivies and feline social relationships. A cat without its claws is not a true cat.

Is it cruel to have a cat de-clawed?

Yes, it is. To remove a cat's claws is far worse than to deprive cat owners of their finger-nails. This is because the claws have so many important functions in the life of a cat. A de-clawed cat is a maimed cat and anyone considering having the operation done to their pet should think again.

Consider the facts. To begin with, it is important that every cat should keep itself well groomed. A smooth, clean coat of fur is essential for a cat's well-being. It is vital for temperature control, for cleanliness, for waterproofing and for controlling the scent-signalling of the feline body. As a result, cats spend a great deal of time every day dealing with their toilet, and in addition to the typical licking movements they perform repeated scratchings. These scratching actions are a crucial part of the cleaning routine, getting rid of skin irritations, dislodging dead hairs and combing out tangles in the fur. Without claws it is impossible for any cat to scratch itself efficiently and the whole grooming pattern suffers as a result. Even if human owners help out with brush and comb, there is no way they can replace the sensitivity of the natural scratching response of their pet. Any people who have ever suffered from itches that they cannot scratch will sympathize with the dilemma of the de-clawed cat.

It has been argued that a de-clawed cat can learn to use its teeth more when grooming. It is true that cats often nibble an irritation, rather than scratch it, but unfortunately some of the most urgent scratching requirements are in the region of the head, mouth, neck and especially the ears. Teeth are useless here and these important parts of the body cannot be kept in perfect condition with only clawless feet to groom them.

A second problem faces the de-clawed cat, when it tries to climb. Climbing is second nature to all small felines and it is virtually impossible for a cat to switch off its urge to climb, even if it is punished for doing so. And punished it certainly will be if it attempts to climb after having its claws removed, for it will no longer have any grip in its feet. Out of doors, if it is being chased by a rival cat, a dog, or some human enemy, it will try, as always, to scamper up a wall or a tree, using its non-existent claws to cling to the surfaces as it leaps upwards. To its horror it will find itself slipping and sliding, tumbling down at the mercy of its foes. When it turns to face them it will be at an even greater disadvantage because, when it strikes out at them with its paws, it will find itself robbed of its defensive weapons and unable to protect itself. Often, it is only the sharpness of the pain caused by the stiletto-pointed claws that stands between life and death for a cornered cat.

In less dramatic contexts, for de-clawed cats kept indoors (and robbed of all outdoor pleasures), even the simple act of climbing up on to a chair or a window-ledge may prove hazardous. Without the pinpoint contact of the tips of the claws, the animals may find themselves slipping and crashing to the ground. The expression of confusion observed on the faces of such cats as they pick themselves up is in itself sufficient to turn any cat-lover against the idea of claw-removal.

In addition to destroying the animal's ability to groom, climb, defend itself against rivals and protect itself from enemies, the operation of de-clawing also eliminates the cat's ability to hunt. This may not be important for a well-fed family pet, but if ever such a cat were to find itself lost or homeless it would rapidly die of starvation. The vital grab at a mouse with sharp claws extended would become a useless gesture.

In short, a de-clawed cat is a crippled, mutilated cat and no excuse can justify the operation. Despite this, many pet cats are carried off to the vet by exasperated owners for this type of 'convenience surgery'. The operation, although nearly always refused by vets in Britain, has become so common in certain countries that it even has an official name. It is called *onyxectomy*. Using an old Greek name for it somehow makes it seem more respectable. The literal translation of onyxectomy, however, is simply 'nail-cutting-out' and that is what vets are doing, even though they may not like to be reminded of the fact when they record their day's work.

It has been argued by some cat experts that any vet performing this convenience operation should be prosecuted for cruelty to animals, but this is unfair to them. They may well be faced with owners who demand that the operation be done, with the only alternative that their pet cat be destroyed. Given such a choice, and a perfectly healthy cat, it is hard to blame the vet for selecting the lesser of two evils. If there is any prosecution, it should be of the owner rather than the vet.

The reason for the popularity of the de-clawing operation in recent years has been the concern of (usually wealthy) owners for their soft furnishings. Valuable chair fabrics, curtains, cushions and other materials are often found scratched, torn and tattered as a result of the family cat's claw-sharpening activities around the house. This is especially troublesome in urban homes, where the animals have little access to wooden posts or trees. And the addition of commercially manufactured 'scratching-posts' to the indoor furniture rarely seems to solve the problem. There are only three solutions: the owners can decide that the pleasure of having a cat in the home is greater than the distress of living with scratched furniture, or some kind of covering that cats hate to scratch has to be put over the sides of the furniture, or the cats have to be trained not to strop the furniture with their claws. This last alternative is a possibility, but it is by no means easy. Shouting at cats or hitting them has little effect. It is best to take a leaf out of the mother cat's book. When she wants to control her kittens, she growls at them. They learn this signal and soon respond to it. Growling like a mother cat may make a cat owner feel ridiculous, but if it helps to restrain a cat from tearing at a valuable chair, it is well worth a try.

Finally, a word of praise for the organizers of pedigree cat shows. All too often they are accused of exploiting cats to satisfy the competitive urges of the exhibitors, and it might be imagined that they would favour the de-clawing operation to facilitate the handling of cats when they are being judged at shows. Nothing could be further from the truth. Despite many a scratched hand, the official ruling is that any pedigree cat found to have had its claws removed is automatically and unconditionally disqualified from competing. Supporting this attitude are new moves in the veterinary world to outlaw totally the operation for any cats, both pedigree and moggie. Britain is leading the way in this and it is hoped that other countries will soon follow suit.

Why do some cats like to take walks with their owners?

Some cat manuals solemnly instruct their readers in how to take their pet cats for a walk. They suggest that you should start the training when your pet is still a kitten and get it used to walking on a harness (not a collar that can be slipped too easily) and lead. You are warned that cats do not take kindly to walking to heel like a dog, and prefer to walk alongside you. Also, long walks, like choke chains, are out of the question.

All this is based on a fundamental misunderstanding about the social life of cats. Adult cats do not go for walks together. They do not explore together, hunt together, flee together or migrate together. When they are on the move they are essentially solitary creatures and find it quite inexplicable that their human owners should wish to indulge in communal walking. If people want to take their pets for walks then they must buy pack animals like dogs, for which group movement is the most natural thing on earth. But cat owners must walk alone.

And yet . . . certain cat owners have reported that *their* cats, unlike others, do like to accompany them on short walks. How can we explain this? If these cases are investigated it is nearly always found that the cats in question are not on leads but are simply following in their owners' footsteps. And these footsteps are nearly always progressing along a well-known garden path or country lane. The village cat owner sets off from home and finds the pet cat tagging along. After a while, the cat gives up and returns to its familiar territory. This is nothing like a synchronized, long-distance, man-and-dog walk and should not be confused with it, but it is a minor walk of sorts and demands some interpretation.

The explanation is found in the behaviour of half-grown kittens. When they have arrived at the fully mobile stage of their development, but have not yet ventured off on their own, they may accompany their mother on short trips away from the 'nest'. She will slow her pace down for them and keep a close eye on them as they amble along near her, but she will not let them get too far away from the home base. When adult cats follow their owners they are reverting to this half-grown kitten stage. For domesticated felines remain 'part-kitten' all through their lives, and even though they may be middle-aged in feline terms they still look upon their human owners as their mothers. So they follow their pseudo-parents as they set off for the village shops, until they feel themselves getting too far from the 'nest', and then break off to return to safety.

Many aspects of adult cat behaviour can be explained in this way, the relics of kittenhood remaining with our pets until they are old and senile.

How have the famous reacted to their feline friends?

Many famous people have owned cats that have played roles in their private lives. Sometimes sensibly and thoughtfully, sometimes romantically and outrageously, they have responded to the presence of their feline companions. The following are just a few examples of the lengths people will go to in their love and respect for their pet cats.

Edward Lear, the Victorian artist and author of *Nonsense Verse*, including the famous story of 'The Owl and the Pussy-cat', adored his tabby cat, Foss. He was so concerned over his cat's comfort that when, late in life, he was moving to a new house in San Remo, he instructed his architects to design his new abode as an exact replica of his old one. This, he felt, would assist the cat to make the transition to the new house with the minimum of disturbance to its feline routine.

Sir Isaac Newton, considered by many to be the greatest natural philosopher of all time, was also concerned over the comfort of his pet cats. He noticed, in particular, that they hated the human invention of the door, which so restricted their free movement in and out of the house. It is somehow apt that, to solve their problem, the great man whose laws of motion and gravity made him world famous should also have invented the cat flap.

Dr Samuel Johnson, the great lexicographer, owned a much-loved and pampered cat named Hodge. His biographer, James Boswell, a confirmed cat hater, was clearly surprised at the trouble Johnson took over his cat, and he records that the great man himself would go out on errands to buy oysters for Hodge, rather than send his staff, 'lest the servants having that trouble should take a dislike to the poor creature'. It seems likely that Johnson was sensitive not only to his staff having to run such errands, but also to the thought that, by doing so, they might start making unfortunate comparisons between their diet and the cat's.

Edgar Allen Poe owned a large tortoiseshell cat named

Catarina. Poe was so poor that he could not afford to provide heat for his sick wife, who was dying of consumption. A visitor reported that the only source of warmth she had as she lay on her straw bed was 'her husband's great coat and a large tortoiseshell cat in her bosom. The wonderful cat seemed conscious of her great usefulness'. In this case, it was the cat that came to the aid of its owners, and its comforting presence inspired the young author to write one of his most horrific stories, 'The Black Cat'. In it, a madman mutilates and kills his cat, only to become haunted by a devil-cat who avenges the pet's death. In other words, he was saying, do not tamper with something as precious as a cat.

Théophile Gautier, the French novelist and poet (who incidentally introduced the phrase 'art for art's sake'), owned a black cat named Eponine, which he had named after a character in Victor Hugo's *Les Miserables*. His regard for the cat was such that whenever he dined alone he always insisted that a place should be set for her at his table. If we are to believe him, she was always promptly in her seat (no doubt with a little help from his servants) at the moment when he entered his dining-room. She sat there, he reported, with 'her paws folded on the tablecloth, her smooth forehead held up to be kissed, and like a well-bred little girl who is politely affectionate to relatives and older people'. Gautier and his cat then proceeded to dine together, the animal first lapping up soup (with some reluctance, he admits) and then feasting on fish.

Jeremy Bentham, the English philosopher who preached that human actions should be aimed at 'producing the greatest happiness for the greatest number' – a concept that has since become the norm for democratic societies – had a much-loved cat named Langbourne. Despite his public concern for 'the greatest number' of people, in private life he seemed not to care for the company of even a small number of them. He was described as 'suffering few persons to visit him, rarely dining out', and was said to consider social activities a waste of his time. Instead, he preferred the company of his adored cat and honoured the animal in a curious way by bestowing titles on him. From simple Langbourne, he became Sir John Langbourne, and was finally awarded a doctorate in divinity by his besotted owner, being given the title of The Reverend Sir John

Langbourne, D.D.

Charles Dickens was plagued by overattentive cats. When his white cat, William, produced a litter of kittens (and was renamed Williamina), she repeatedly insisted on carrying each kitten from its official home in the kitchen and placing it in the master's study. After he had evicted them several times, in order to concentrate on his writing, she returned once more and dropped them one by one at his feet. At this point, he gave in. Later, his writing suffered once again when one of the grown-up kittens he had kept and christened The Master's Cat developed a remarkable habit of snuffing out his candle with her paw as a way of diverting his attention away from his books and towards her.

The prophet Mohammed is said to have loved his cat, Meuzza, so much that he sacrificed his favourite robe to the animal. Meuzza was asleep on the sleeve of this robe one day when Mohammed was summoned to prayer and, rather than disturb the cat's serene slumber, he cut off the sleeve and left the animal snoozing peacefully. After he returned and the cat awoke, Mohammed stroked the animal three times and Meuzza was thus given a permanent place in the Islamic Paradise.

Ichijo, an early Japanese emperor, was so enamoured of his cat that when Myobo No Omoto (which means Omoto, lady-in-waiting) was chased by a dog he had the unfortunate canine exiled and its human companion imprisoned.

When Alexander Dumas's cat, Mysouff II, ate his entire collection of exotic birds, instead of banishing the hungry feline, he gave it a fair trial before his guests the following Sunday. One guest pleaded the animal's defence on the grounds that the aviary door had been opened by Dumas's pet monkeys, and this was considered to be 'extenuating circum-stances'. The sentence agreed upon was that the unfortunate cat should serve five years' imprisonment in the monkeys' cage. Luckily for Mysouff II, however, Dumas was shortly to find himself financially embarrassed. As a result, he had to sell the monkeys, and the cat regained its freedom.

Karel Capek, the Czech playwright, felt himself to be magically infested with cats. On the very day his Angora tomcat died of poison, an avenging female cat appeared on his

201

doorstep. Its mission, he mused, was to 'revenge and replace a hundredfold the life of that tomcat'. Christened Pulenka, she set about this task with a reproductive verve that staggered him, producing in a very short space of time no fewer than twenty-six kittens. One of her daughters, Pulenka II, continued with the 'plot', presenting Capek with twenty-one more kittens before she was killed by a dog. One of *her* daughters, named Pulenka III, in turn continued what he called 'The Great Task' of creating a host of cats to seize power 'to rule over the universe'.

Robert Southey, the English author and poet who was an early figure in the Romantic movement, had such regard for his cat that he insisted on giving it a formal name that makes even the most elaborate pedigreed name seem economical. The cat's full title was The Most Noble the Archduke Rumpelstizchen, Marquis Macbum, Earle Tomemange, Baron Raticide, Waowler, and Skaratchi. Known as Rumpel to his friends, his death caused the whole household, servants included, to mourn his passing, Southey remarking that the sense of bereavement was greater than any of them liked to admit.

Theodore Roosevelt owned a stubbornly independent cat named Slippers. The animal often disappeared from the White House, only to return at inopportune moments. On one occasion, the cat had come home and sprawled itself unexpectedly across a rug in the middle of a main corridor. Roosevelt, with an ambassador's wife on his arm, appeared, leading the formal procession of important guests attending a state banquet. Confronted by Slippers' recumbent form, the President was faced with a difficult choice: disturb his favourite, pernickity old cat, or divert the whole procession. In the great tradition of cat lovers, he chose the latter course. Bowing to his partner, he led her around the cat's body and the entire procession of dignitaries followed suit, studiously ignored by the regally reclining cat.

The attitudes of all these famous cat owners can best be summed up by the words of Mark Twain: 'A house without a cat, and a well-fed, well-petted and properly revered cat, may be a perfect house, perhaps, but how can it prove its title?'

Why are cat-owners healthier than other people?

This may sound a strange question, but there is a great deal of evidence to prove that cat-owning is good for your health. And there is some comfort for beleaguered pet-owners, often criticized today for 'messing up the environment with their animals', that the anti-pet lobby will die younger than they will.

There are two reasons for this. First, it is known that the friendly physical contact with cats actively reduces stress in their human companions. The relationship between human and cat is touching in both senses of the word. The cat rubs against its owner's body and the owner strokes and fondles the cat's fur. If such owners are wired up in the laboratory to test their physiological responses, it is found that their body systems become markedly calmer when they start stroking their cats. Their tension eases and their bodies relax. This form of feline therapy has been proved in practice in a number of acute cases where mental patients have improved amazingly after being allowed the company of pet cats.

We all feel somehow released by the simple, honest relationship with the cat. This is the second reason for the cat's beneficial impact on humans. It is not merely a matter of touch, important as that may be. It is also a matter of psychological relationship which lacks the complexities, betrayals and contradictions of human relationships. We are all hurt by certain human relationships from time to time, some of us acutely, others more trivially. Those with severe mental scars may find it hard to trust again. For them, a bond with a cat can provide rewards so great that it may even give them back their faith in human relations, destroy their cynicism and their suspicion and heal their hidden scars. And a special study in the United States has recently revealed that, for those whose stress has led to heart trouble, the owning of a cat may literally make the difference between life and death, reducing blood pressure and calming the overworked heart.

What are the games cats play?

One of the great pleasures of owning a cat is watching it at play. Or playing with it oneself. For hundreds of years this innocent diversion has fascinated even the most learned of men. As long ago as the second century A.D., the Roman historian Lucius Coelius recorded that, when he was free from his studies and more weighty affairs, he was not ashamed to play and sport himself with his cat. The great naturalist Edward Topsel, in the seventeenth century, waxed lyrical about feline play: 'Therefore how she beggeth, playeth, leapeth, looketh, catcheth, tosseth with her foot, riseth up to strings held over her head, sometimes creeping, sometimes lying on the back, playing with one foot, sometimes on the belly, snatching now with mouth, and anon with the foot . . .' Concerned, however, that these words might make him seem too frivolous, in the simple delight he obviously took in playing with his cat, he checks himself with the criticism that 'verily it may well be called an idle man's pastime'. Warming to this thought he then goes on to castigate cat-lovers 'because they which love any beast in a high measure, have so much less charity unto man.'

Michel de Montaigne, a French writer of the same era, managed to avoid this hypocrisy, stating honestly that, 'When I play with my cat who knows whether she diverts herself with me, or I with her. We entertain one another with mutual follies . . . and if I have my time to begin to refuse, she also has hers.' These remarks reveal the irresistible appeal of the playful cat even in an era when felines were generally considered to be evil and dangerous, and were being widely persecuted. Topsel warns that playing with them may destroy the lungs and corrupt the air: 'There was a certain company of Monks much given to nourish and play with Cats, whereby they were so infected, that within a short space none of them were able to say, read, pray, or sing, in all the Monastery . . . ' But despite these ridiculous fears and superstitions, the cat at play continued to weave its magic and enchant all but the most hypochondriac observers.

Some, however, were concerned about the nature of the play activities, almost all of which were seen to be based on some kind of violence. Even the smallest kittens, they noticed, were acting out attacks either on other members of their own species or on prey animals. In the eighteenth century the French naturalist Buffon remarked: 'Young cats are gay, vivacious, frolicksome, and, if nothing was to be apprehended from their claws, would afford excellent amusement to children. But their toying, though always light and agreeable, is never altogether innocent, and is soon converted into habitual malice.' This is a second form of hypocrisy. Not only does Buffon try to relegate the pleasure of watching kittens at play to the level of a childish amusement, but he also suggests that this play is malicious because it involves acting out the killing of prey — the very skill for which mankind domesticated the cat in the first place.

These moral issues, which troubled those who wished to enjoy the company of animals but felt they had to judge them on a set of human values, have caused less concern in recent years. The gradual acceptance of the Darwinian evolutionary philosophy, that enables us to accept each species in its own right and see its actions from its own point of view, has meant that we are released from the burden of interpreting everything animals do in terms of good and evil. If a cat kills a mouse we may feel sorry for the mouse but we no longer accuse the cat of wickedness. Equally, if the mouse was a pest and we are pleased to see the end of it, we do not praise the cat for the saintly way in which it has fulfilled its duty to its household. Our whole attitude has changed. We now see the cat's actions as part of its natural specialization as a predatory carnivore, and we recognize that in ridding our houses of mice it is neither 'malicious' and 'vicious' on the one hand, nor 'loyal' and 'responsible' on the other. It is merely a cat being a cat.

With this new approach we are able to sit back and enjoy the delightful escapades of young kittens at play without any moral posturing. We are able to become objective observers of the patterns of play and to take pleasure in their endless variations on central themes. Four of these have been identified.

The earliest to develop is that of play-fighting. At about three weeks the kittens start to engage in rough-and-tumble actions with their litter-mates. They jump on one another, roll over on their backs and grapple. No one gets hurt. This is because at first they lack the strength to hurt, and when they do acquire it they quickly learn that a too-powerful play-attack ends the enjoyable encounter. So they perfect the art of inhibited assault. By the age of four weeks the play-fighting becomes more elaborate, with chasing, pouncing, clasping with the front legs and vigorous kicking with the hind legs. And now the other main play-themes are added, each connected with a different kind of prey-hunting. They have been given the names of 'the mouse-pounce', 'the bird-swat' and 'the fish-scoop'.

The mouse-pounce involves hiding, crouching, creeping forward, and then rushing and pouncing on an imaginary rodent, usually its mother's twitching tail or a small object lying on the ground. The bird-swat includes the same approach, but then ends with an upward leap and a sharp blow with the front foot. The stimuli that trigger this action are usually moving objects that hang down from above, or toys that are thrown to the kittens by their owners. The fish-scoop occurs when the object lying on the ground is very static. The kitten suddenly flings out a paw and scoops the object up into the air and backwards over its shoulder. It then turns and pounces on it triumphantly, as if a fish scooped up from a river or stream has been landed on the bank and must be secured before it wriggles its way back to the safety of the water.

During these play bouts the kitten's imagination is put to full use. Anything small that moves easily may be accepted as a victim. Expensive toys are available in the form of wind-up mice, tinkling cylinders and catnip-soaked balls, and both young kittens and older cats may react to these with great intensity. This interest, however, may be rather brief, for these toys usually lack two essential qualities. They are too hard and too heavy. The ideal toy is very light, so that only a small amount of effort moves it a long way, and very soft, so that sharp feline claws and teeth can sink into it in a satisfying way. Ironically, this means that the most exciting objects for play are also the simplest and the cheapest. A piece of silver wrapping-paper rolled up into a tight ball, or the traditional ball of wool, provides the greatest reward. One owner after another makes the amused discovery that these simple objects will keep the playful cat or kitten occupied much longer than any fancy toy.

All cats show the four basic patterns of play described here, but in addition each pet feline may develop its own special games. These become almost like rituals as the cat grows older – little routines that reward the animal because they involve it in a social interaction with its owners or their guests. Thomas Huxley, the great biologist, whose household was dominated by a long series of cats over a period of forty years, described how one of them, a young tabby tom-cat, developed the alarming game of jumping on the shoulders of his dinner-guests and refusing to dismount until they fed him some titbit. It was not that the animal was hungry. It was the shock impact of the game that provided the reward.

One owner discovered another routine. If he put down sheets of newspaper in the kitchen to keep the floor clean on wet days, one of his cats would back up to the far wall and then launch herself as fast as she could at the papers. The moment she hit one of the papers, she braked and went into a long skid, sliding right across the kitchen floor to the other wall, which she thumped into, still standing on her 'magic carpet'. After this she would return to the far wall and wait for the papers to be put back into place, so she could repeat the game.

Another owner discovered that if he put a line of coins on his sideboard his cat would knock them down one by one. Eventually he managed to turn this into a special trick, with the cat knocking down a coin each time his owner clicked his fingers. The same cat enjoyed jumping from chair to chair when his owner pointed at each in turn.

The more one talks to individual cat owners, the greater the variety of cat personalities one finds. It is true that all cats share many features of their behaviour, down to the tiniest detail. But when it comes to playtime, each cat seems to have its own personal, idiosyncratic way of embellishing its playful interactions with its owners. If it is lucky, it will have owners who are of an equally playful frame of mind.

Which are the most expensive cats?

When a pet cat produces a litter of kittens, its owners usually have a hard time finding volunteers to take the young ones. For most people, it is easy enough to acquire a kitten without money changing hands. The animal shelters are full of unwanted felines waiting for suitable homes, and in addition, there are stray cats wandering suburbia in search of a soft-hearted human to take them in. But some people want a special cat and are prepared to pay a great deal of money for it.

Kittens with good pedigrees that might become winners of cat championships fetch high prices. Today such a kitten will cost a would-be competitor over a hundred pounds. A proven champion may fetch much more. One American breeder wanted a champion white Persian tomcat so badly that the British owner was offered about $3,500. That was many years ago. If the offer had been made today, the equivalent sum would be about $33,000, which probably constitutes a record bid for any cat.

The most expensive pet cats, as opposed to championship cats, are those offered for sale recently by the luxury New York department store, Neiman-Marcus. Described as 'Designer Kittens', they are offered in any colour clients may desire to match their clothes or their house decorations. The store claims that these exclusive animals are the creation of a geneticist who has been deliberately mixing up a variety of pure pedigreed cats to produce a whole range of colours to suit every taste, from red, gold, silver, and bronze to charcoal. In addition to its special colouring, each kitten boasts the 'markings of a jungle cat' to make it stand out from any ordinary pet cat. The cost of one of these specially bred kittens is a cool $1,400.

Designer Kittens produced an outcry when they first surfaced in 1986. Animal welfare groups were outraged, complaining that cats should not be marketed as 'high-priced toys', and they warned would-be purchasers that 'crossbreeding can produce cats with strange dispositions'. They cautioned them that the animals' behaviour, when they matured into full-grown cats, might surprise them. This is a nonsensical criticism and is the kind of attack that brings animal welfare organizations into disrepute. The simple truth is that they find the idea of people paying such huge sums of money for kittens repulsive, when they themselves have to kill perfectly healthy but unwanted kittens every day. A more honest statement along these lines would be much more effective than inventing imaginary mental conditions for crossbred cats. In reality, crossbreeding produces hybrid vigour and creates cats that are sturdier and live longer than purebred lines. If a geneticist has been tinkering with the colour patterns of pedigreed cats to produce some attractive new 'designs', there is nothing inherently wrong in that, and if people wish to pay huge sums of money to own an exclusive 'designer' cat, there is nothing inherently wicked about that either. At those prices, the animals are more likely to be well looked after than are many of the unwanted give-away kittens.

In the end, the price of the cat does not matter, nor does the colour of its coat. All that matters is that it finds itself in a friendly, caring household, and it is just as likely to do this at the top end of the market as at the bottom.

Why do so many black cats have a few white hairs?

Anyone owning an ordinary black moggie, as distinct from a pedigree 'Black Shorthair', may well have noticed that it boasts a small patch of white hairs, sometimes boldly visible and sometimes barely perceptible. Occasionally an otherwise totally black cat may have a single white whisker. More commonly, there is a touch of white on the chest. Why should this be?

It is certainly no accident. In fact it is a remnant of a disastrous period in the history of European felines. The origins of the black cat, as a distinct colour type, have been traced back to the ancient Phoenicians, who sneaked some of the sacred cats out of Egypt and began trading in them around the Mediterranean. During their travels they appear to have developed a black variety which became extremely popular, perhaps because its natural nocturnal camouflage assisted it to become a more efficient mouser and ratter. The black cat spread all over Europe until, in medieval times, it became associated with black magic and sorcery. For several centuries after this it was persecuted. The Christian Church organized annual burning-cats-alive ceremonies on the day of the Feast of St John. For these cruel rituals the most wicked and depraved of 'Satan's felines' were strongly preferred and all-black cats were eagerly sought out for the flames. But these cats had to be totally black to be really evil, in the minds of the pious worshippers. Any touch of white on their black coats might be taken as a sign that they were not, after all, cats consecrated to the Devil.

As a result of this distinction, cats that were totally black became less and less common, while those that were black with a touch of white survived. Religion acted as a powerful selection pressure on feline coloration.

By the seventeenth century, when this persecution of cats was beginning to wane, a new danger arose. It was now believed that they – or vital parts of them – were infallible cures for a whole variety of ailments and weaknesses. The tail of an all-black cat, severed and buried under the doorstep of a house, was considered to be a way of preventing all members of the family living there from succumbing to sickness and ill health.

Edward Topsel, the English naturalist, writing in 1658 stipulated that, to cure blindness, or pains in the eye: 'Take the head of a black Cat, *which hath not a spot of another colour in it*, and burn it to powder in an earthen pot leaded or glazed within, then take this powder and through a quill blow it thrice a day into the eye' – the italics were not used in the original, but are to draw attention to the crucial quality of the black cat who is about to lose his head. So, after many years of religious selection pressure working against the pure-black cat, there was now the added pressure of medical quackery. Little wonder that today's black moggies so frequently sprout a small patch of white hairs as a badge of protection from earlier human follies.

Today a pure-black cat is probably a special pedigree specimen. Since competitive cat shows began, there has been a third selection pressure operating on feline coat colours – that of purification. Now, any kitten with tufts or streaks of white hair on an otherwise jet black coat will be ignored. Only those without such adornments – once so vital – will be selected for further breeding. Even so, it is significant that in most of the volumes on cat breeding and pedigree standards, there is a small, tell-tale phrase under the heading of 'coat colours', which states that the fur 'should be dense with no trace of white hairs'. The need to make such a comment for a pedigree breed that is clearly stated to be all-black in colour is a powerful reminder that the little white patches and flashes are still persisting, even after a century of specialized, selective breeding.

For those of us with non-pedigree black moggies which sprout a few white hairs, it is comforting to think of their special markings not as some kind of mongrel flaw but instead as a vital and valuable relic of earlier days in the feline history of Europe.

How did the cat become associated with witchcraft?

Religious bigots have often employed the cunning device of converting other people's heroes into villains, to suit their own purposes. In this way, the ancient horned god that protected earlier cultures was first transformed into the evil Devil of Christianity. And the revered, sacred feline of ancient Egypt became the wicked, sorcerer's cat of medieval Europe. Anything considered holy by a previous religious faith must automatically be damned by a new religion. In this way began the darkest chapter in the cat's long association with mankind. For centuries it was persecuted and the cruelties heaped upon it were given the full backing of the Church.

During this bleak phase of its history the cat became firmly linked in the popular mind with witches and black magic. As late as 1658 Edward Topsel, in his serious work on natural history, followed detailed descriptions of the cat's anatomy and behaviour with the solemn comment that 'the familiars of Witches do most ordinarily appear in the shape of Cats, which is an argument that this beast is dangerous to soul and body.'

Because the cat was seen as evil, all kinds of frightening powers were attributed to it by the writers of the day. Its teeth were said to be venomous, its flesh poisonous, its hair lethal (causing suffocation if a few were accidentally swallowed), and its breath infectious, destroying human lungs and causing consumption. All this created something of a problem, since it was also recognized that cats were useful 'for the suppressing of small vermine'. Topsel's compromise was to suggest to his readers that 'with a wary and discreet eye we must avoid their harms, making more account of their use than of their persons'. In other words, exploit them, but do not get too close to them or show them any affection.

This restrained attitude did, at least, enable farm cats and some town cats to live a tolerable life as unloved pest-controllers, but for certain village cats life was far more unpleasant. If they happened to attach themselves to an old woman who lived by herself, they were risking a savage death as a witch's familiar. The sad irony of this unhappy state of affairs was that these animals played a comforting role in the lives of such old women. Any elderly crone, who happened to be ugly or misshapen enough to have repelled all potential husbands, and who was therefore forced to live a solitary life with no children of her own, often as an outcast on the edge of the village, was desperately in need of companionship. Maltreated cats, finding themselves in a similar plight, often approached such women, who befriended them as substitutes for human companionship and love. Together, they brought one another many rewards and the kindness of these old women towards their cats was excessive. Anyone teasing or hurting their beloved felines was cursed and threatened. All that was required then was for one of these tormentors to fall ill or suffer a sudden accident and the old 'witch' was to blame. Because the cats wandered about, often at night, they were thought to be either the supernatural servants of the witches, or else the witches themselves, transformed into cat-shape to aid their nocturnal travels when seeking revenge.

So it was the reaction against the cat's ancient 'holiness' in Egyptian religion, combined with its connection with childless, elderly women, that made it the 'wicked' animal of the medieval period. Added to this was its haughtiness and its refusal to become completely subservient to human demands, unlike the dog, the horse, the sheep and other easily controlled domestic animals. Also, its nocturnal caterwauling during the breeding season gave rise to tales of orgies and secret feline ceremonies. The outcome was a savage persecution lasting for several centuries, perpetrated against an animal whose only serious task was to rid human habitations of infestations of disease-carrying, food-spoiling rats and mice. It is a strange chapter in the history of Christian kindness.

Why does a black cat bring good luck?

In Britain it has long been believed that if a black cat crosses your path or enters your house it will bring you good fortune. This superstition has three origins.

The first source takes us back to ancient Egypt where the sacred cat was thought to bestow many blessings on the household that looked after it. Its real blessing, of course, was that it kept down the numbers of mice and rats. But this was extended, through myth and legend, to be seen as a generalized blessing. Ancient tomb inscriptions inform us that the cat 'gives life, prosperity and health every day, and long life and beautiful old age'. Clearly no home should be without one. By slender threads, this ancient tradition managed to cling on, century after century, even when the Christian Church started its onslaught against cats as the servants of the Devil.

The second source is from medieval times, when the 'devil-cat' was feared and hated. It was then believed that if a cat crossed your path and did you no harm, you had been incredibly lucky. Hence the association between cats and good luck. If one came into your house then, by being kind to it, you could appease its master, the Devil, and avoid his wrath. So a cat entering your house and being welcomed there gave you the good luck of having the Devil on your side. Others might be tormented by him, but not you.

The third source is more down to earth. An old British saying is: 'Whenever the cat of the house is black, the lasses of lovers will have no lack.' Here the cat is being used as a symbol of sexual attraction. The female cat on heat attracts a whole circle of admiring tom-cats, and the house with such a cat is therefore a place where perhaps any female, even human, would be successful at attracting a large circle of male admirers.

The colour black in all such cases was considered especially lucky because this was the colour associated with the occult practices, but herein lies a transatlantic contradiction for, in

America, it is the white cat that is lucky and the black cat that is unlucky. There it seems that from the earliest days of the pioneer settlers the black cat was linked with the devil so strongly that it was, in any context, an evil force. Nobody was prepared to have any dealings with this force, even to attempt to pacify it by being kind to the black cat. The white cat, presumably by direct contrast, was seen as a force of light against the darkness and was in this way converted into a symbol of good fortune.

Why do we speak of not having a 'cat-in-hell's' chance?

At first sight this is as puzzling as the well-known footballer's lament of being as 'sick as a parrot'. In both cases the mystery is solved if you know the original, unabbreviated saying which has long since been discontinued. The complete cat phrase is: 'No more chance than a cat in hell without claws'. It was originally a reference to the hopelessness of being without adequate weapons. (The original parrot saying, incidentally, was 'as sick as a parrot with a rubber beak' – a similar allusion to the lack of a sharp weapon.)

How has the cat been used in warfare?

Although it is common knowledge that dogs have been extensively employed in times of war, most people imagine that the cat has never been exploited in this way. But they are wrong. There are two examples, albeit somewhat unusual ones.

The first dates back two and a half thousand years to the time when the Persians were at war with the Egyptians. Knowing that the Egyptians revered the cat and considered it to be sacred, the Persians developed the idea of a 'feline armour'. When their advance guard was making a hazardous push to secure a new stronghold, the Persian warriors went forward carrying live cats in their arms. Seeing this, the Egyptian soldiers were unable to attack them in case they accidentally injured or killed one of these sacred animals. For them, such act of violence against one of their animal deities was unthinkable. Indeed, if any one of them had killed a cat, even in these special circumstances, he would have been put to death for it. So in this way the Persians were able to advance with ease and the Egyptians were helpless to retaliate.

Incidentally, despite these defeats in war, the Egyptians did not weaken in their worship of the cat. We know from the observations of Herodotus nearly a century later that they were still treating the animal with the utmost respect. He reported that if a house caught fire nobody attempted to put it out because they were all concentrating on protecting the cats, forming up in lines to prevent the panic-stricken animals from running into the flames and burning themselves. And whenever a cat died he noted that Egyptians all went into deep mourning and shaved off their eyebrows as a sign of their distress. So it is little wonder that the Persians found the cat an invaluable ally even at the height of battle.

The second example of the use of cats in warfare appears much later, being illustrated in Christopher of Hapsburg's book in the year 1535. He was an artillery officer and in his report to the Council of One and Twenty at Strasbourg he described the way in which 'poisoned vapours were shed abroad' by means of cats. The unfortunate animals apparently had poison bottles strapped to their backs, with the openings pointing towards their tails. They were then sent off towards the enemy, running panic-stricken this way and that, and in the process spreading the poisonous fumes. Christopher of Hapsburg was clearly a man of delicate sensibilities, for he adds the comment, 'This process ought not to be directed against Christians.'

Why do we say 'there is no room to swing a cat'?

This refers to the whip called 'the cat', employed on early naval vessels, and not to the animal itself. The cat, or cat-o'-nine-tails (because it has nine separate knotted thongs), was too long to swing below decks. As a result, sailors condemned to be punished with a whipping had to be taken up above, where there *was* room to swing a cat.

The reason why the whip itself was called a cat was because it left scars on the backs of the whipped sailors reminiscent of the claw marks of a savage cat.

Why do we talk about catnaps?

Because cats indulge in brief periods of light sleep so frequently. In fact, these short naps are so common in cats and so rare in healthy humans that it is not exaggerating to say that cats and people have a fundamentally different sleep pattern. Unless human adults have been kept awake half the night, or are sick or extremely elderly, they do not indulge in brief naps. They limit their sleeping time to a single prolonged period of approximately eight hours each night. By comparison, cats are super-sleepers, clocking up a total, in twenty-four hours, of about sixteen hours, or twice the human period. This means that a nine-year-old cat approaching the end of its life has only been awake for a total of about three years. This is not the case in most other mammals and puts the cat into a special category – that of the refined killer. The cat is so efficient at obtaining its highly nutritious food that it has evolved time to spare, using this time to sleep and, apparently, to dream. Other carnivores, such as dogs and mongooses, have to spend much more time scurrying round, searching and chasing. The cat sits and waits, stalks a little, kills and eats, and then dozes off like a well-fed gourmet. Nothing falls asleep quite so easily as a cat.

There are three types of feline sleep: the brief nap, the longer light sleep and the deep sleep. The light sleep and the deep sleep alternate in characteristic bouts. When the animal settles down for more than a nap, it floats off into a phase of light sleep which lasts for about half an hour. Then the cat sinks further into slumber and, for six to seven minutes, experiences deep sleep. After this it returns to another bout of thirty minutes of light sleep, and so on until it eventually wakes up. During the periods of deep sleep the cat's body relaxes so much that it usually rolls over on to its side and this is the time when it appears to be dreaming, with frequent twitchings and quivering of ears, paws and tail. The mouth may make sucking movements and there are even occasional vocalizations, such as growls, purrs and general mutterings. There are also bursts of rapid eye movement, but throughout all this the cat's trunk remains immobile and totally relaxed. At the start of its life, as a very young kitten during its first month, it experiences only this deep type of sleep which lasts for a total of about twelve hours out of every twenty-four. After the first month the kittens rapidly switch to the adult pattern.

Why is a female cat called a queen?

Because when she is on heat she lords it over the toms. They must gather around her like a circle of courtiers, must approach her with great deference, and are often punished by her in an autocratic manner.

Why is a male cat called a tom?

This can be traced back precisely to the year 1760 when an anonymous story was published called *The Life and Adventures of a Cat*. In it the 'ram cat', as a male was then known, was given the name 'Tom the Cat'. The story enjoyed great popularity, and before long anyone referring to a male cat, instead of calling it a 'ram', used the word Tom, which has survived now for over 200 years.

Why is a cat called a cat?

We take the names of our animals for granted, but if we trace them back to their roots they often tell us something about the origins of the animals themselves. So why is a cat a cat?

The name 'cat' is used by almost every European nation with a slight variation: in French, *chat*; German, *Katze*; Italian, *gatto*; Spanish, *gato*; Swedish, *katt*; Norwegian, *katt*; Dutch, *Kat*; Icelandic, *kottur*; Polish, *kot*. It is also found in countries around the Mediterranean: in Yiddish it is *kats*, in Greek, *gata*, and in Maltese *qattus*. Clearly, this is an ancient word that has spread across the world from a single source. The source appears to be Arabic, because the oldest use of it is found in North Africa, where it is *quttah*. And there is a similar word used by the Berber tribesmen.

This fits with the idea that all domestic cats are descended from the North African wild cat, *Felis lybica*, via domestication by the ancient Egyptians. The Egyptians also provide us with an explanation of why we call a cat 'puss' or 'pussy', these being variations of the name of the early Egyptian cat goddess, *Pasht*. Tying the animal even tighter to this part of the world is the origin of the word 'tabby', which comes from its Turkish name *utabi*. And the general word for a cat in Turkey is *kedi*, which has probably given us our pet word 'kitty'.

What is the origin of cat's cradle?

Cat's cradle is a game played by children in which a loop of string is wound back and forth over the fingers of both hands to create a pattern. Once the first child has created a pattern, the loop is transferred to the hands of a second child, creating a new pattern in the process. The loop is then handed back and forth repeatedly, making a different pattern each time. The starting pattern is called the 'cradle', and the simplest explanation for this is that the cradlelike form is just about the right size for a cat. But this interpretation leaves something to be desired. Why should a cat have been chosen to give the game its name? Why not some other small animal? A more ancient and satisfying explanation has been sought, and two rival solutions have been found.

The first relates to an Eastern European custom. It used to be believed that the cat could increase the chances of fertility in a young married couple. This was presumably based on the observation that cats themselves so seldom have any difficulty in producing large numbers of offspring. The cat had a special role in a fertility ritual performed one month after a wedding had taken place. A cat was secured in a cradle, and this was then ceremonially carried into the newlyweds' house, where it was rocked back and forth in their presence. This, it was claimed, would ensure an early pregnancy for the young bride.

From this ceremony it is easy to see how the idea of a cat's cradle could have entered the folklore of children's games and then spread right across Europe and even to the New World, with the original significance soon forgotten.

Students of ancient Egypt have an entirely different explanation, however. They point out that similar string games are played by peoples as far apart as Congo tribesmen and the Eskimos, and that these games have a magical significance. The string patterns are formed, altered and re-formed in the belief that these actions will influence the path of the sun. In the Congo this is done to persuade the sun to rest; in the frozen north it is just the opposite – the Eskimos try to trap the sun in their strings to shorten its winter absence. The sun in these cases is envisaged as a 'solar cat', to be symbolically ensnared in the twisting string patterns. If this seems farfetched, it should be remembered that in the civilization of early Egypt the great sun god Re (or Ra) was thought to take the form of a cat in his battle with the powers of darkness, symbolized by a giant serpent. At dawn each day, the valiant sun-cat cut off the head of the serpent, defeating the night and bringing the light of another day. This equation between the cat and the sun is thought by some to have spread across the globe from culture to culture and to provide the true origin of the magical game of cat's cradle.

Of the two explanations, the fertility ceremony, with a cat inside a cradle, seems to be the more plausible, although the early Egyptian legend may even have had an influence in the origin of this, too.

Why is a brothel called a cathouse?

Prostitutes have been called cats since the fifteenth century, for the simple reason that the urban female cat attracts many toms when she is on heat and mates with them one after the other. As early as 1401, men were warned of the risks of chasing the 'cattis tailis', or cat's tail. This also explains why the word 'tail' is sometimes used today as slang for female genitals. A similar use for the word 'pussy' dates from the seventeenth century.

What is catgut?

Despite its name, it is not the guts of a cat. It comes instead from the entrails of sheep. Their intestines are cleaned, soaked, scraped, and then steeped for some time in an alkaline liquid. After this they are drawn out, bleached, dyed, and twisted into cords. These cords have great strength and flexibility and have been used for centuries in the making of stringed musical instruments. In earlier days they were also employed as bowstrings for archers.

In this case, why should sheepgut be perversely referred to as catgut? The clue lies in the earliest use of the term. At the beginning of the seventeenth century, one author wrote of fiddlers 'tickling the dryed gutts of a mewing cat'. Later we heard of a man upset 'at every twang of the cat-gut, as if he heard at the moment the wailings of the helpless animal that had been sacrificed to harmony'. These references come from a period when domestic cats were all too often the victims of persecution and torture, and the sound of squealing cats was not unfamiliar to human ears. In addition, there was the noise of the caterwauling at times when feral tomcats were arguing over females in heat. Together, these characteristic feline sounds provided the obvious basis for a comparison with the din created by inexpert musicians scraping at their stringed instruments. In the imaginations of the tormented listeners, the inappropriate sheepgut became transformed into the appropriate catgut – a vivid fiction to replace a dull fact.

Why is a non-pedigree cat called a moggie?

A non-pedigree dog is always referred to as a mongrel and, strictly speaking, this is the correct term for a non-pedigree cat, but few people use it in this way. They are much more likely to call their pet feline a 'moggie' (sometimes spelt moggy).

Despite its popularity, few people seem to know the origin of the term. It began life as a local dialect variant of the name 'Maggie' and its original meaning was 'a dishevelled old woman'. In some regions it was also the name given to a scarecrow and the essence of its meaning was that something was scruffy and untidy. By the start of the present century its use had spread to include cats. This seems to have begun in London where there were countless scruffy alley-cats whose poor condition doubtless led to the comparison with 'dishevelled old women'.

By the inter-war period the word moggie had been abbreviated to 'mog' and in the 1920s and 1930s schoolboy slang referred to dogs and cats as 'tikes and mogs'. For some reason, this shortened form fell into disuse after the Second World War and the more affectionate 'moggie' returned as the popular term for the ordinary, common-or-garden cat.

Why do we say that someone is grinning like a Cheshire Cat?

In Lewis Carroll's *Alice in Wonderland* we encounter a large cat, lying on the hearth and grinning from ear to ear. Alice is told that the reason it is grinning is because it is a Cheshire Cat. There is, however, no explanation as to why cats from that particular English county should be prone to smiling. A clue comes with the final disappearance of the cat, when it slowly vanishes, starting with the end of its tail and ending with the broad grin, which remains some time after the rest of the animal has gone. It is this disembodied grin that explains the source of Lewis Carroll's image, for there used to be a special kind of Cheshire cheese which had a grinning feline face marked on one end of it. The rest of the cat was omitted by the cheesemaker, giving the impression that all but the grin had vanished.

Lewis Carroll may well have seen these cheeses. But he may have taken his reference from an even earlier source. The reason why the Cheshire cheesemakers saw fit to add a grinning cat to their product was because the expression to 'grin like a Cheshire Cat' was already in use for another reason altogether. It was an abbreviation of the saying to 'grin like a Cheshire Caterling', which was current about five centuries ago. Caterling was a lethal swordsman in the time of Richard III, a protector of the Royal Forests who was renowned for his evil grin, a grin that became even broader when he was despatching a poacher with his trusty sword. Caterling soon became shortened to 'Cat' and anyone adopting a particularly wicked smile was said to be 'grinning like a Cheshire Cat'. Lewis Carroll possibly knew of this phrase but, because he refers to the grin outlasting the rest of the body, it is more likely that his real influence was the cheese rather than the swordsman.

Whichever is the case, the fact remains that the saying does not start with Carroll, as most people assume, but was in reality much older and was merely borrowed and made more famous by him.

225

Why do we speak of someone 'having kittens'?

When we say 'she will have kittens if she finds out about this' we mean that someone will be terribly upset, almost to the point of hysteria. At first sight there is no obvious connection between distraught human behaviour and giving birth to kittens. True, a panic-stricken or hysterical woman who happens to be pregnant might suffer a miscarriage as a result of the intense emotional distress, so suddenly giving birth as a result of panic is not hard to understand. But why kittens? Why not puppies, or some other animal image?

To find the answer we have to turn the clock back to medieval times, when cats were thought of as the witch's familiars. If a pregnant woman was suffering agonizing pains, it was believed that she was bewitched and that she had kittens clawing at her inside her womb. Because witches had control over cats, they could provide magical potions to destroy the litter, so that the wretched woman would not give birth to kittens. As late as the seventeenth century an excuse for obtaining an abortion was given in court as removing 'cats in the belly'.

Since any woman believing herself to be bewitched and about to give birth to a litter of kittens would become hysterical with fear and disgust, it is easy to see how the phrase 'having kittens' has come to stand for a state of angry panic.

Why do we say 'it is raining cats and dogs'?

This phrase became popular several centuries ago at a time when the streets of towns and cities were narrow, filthy and had poor drainage. Unusually heavy storms produced torrential flooding which drowned large numbers of the half-starved cats and dogs that foraged there. After a downpour was over, people would emerge from their houses to find the corpses of these unfortunate animals, and the more gullible among them believed that the bodies must have fallen from the sky – and that it had literally been raining cats and dogs.

A description of the impact of a severe city storm, written by Jonathan Swift in 1710, supports this view: 'Now from all parts the swelling kennels flow, and bear their trophies with them as they go ... drowned puppies, stinking sprats, all drenched in mud, dead cats, and turnip-tops, come tumbling down the flood.'

Some classicists prefer a more ancient explanation, suggesting that the phrase is derived from the Greek word for a waterfall: *catadupa*. If rain fell in torrents – like a waterfall – then the saying 'raining catadupa' could gradually have been converted into 'raining cats and dogs'.

Why does a cat have nine lives?

The cat's resilience and toughness led to the idea that it had more than one life, but the reason for endowing it with nine lives, rather than any other number, has often puzzled people. The answer is simple enough. In ancient times nine was considered a particularly lucky number because it was a 'trinity of trinities' and therefore ideally suited for the 'lucky' cat.

Why is the cat so popular?

For thousands of years the domestic cat has aroused strong emotions in its human companions. In ancient Egypt it was adored to the point of worship, and cat-killing was punishable by death. In the Dark Ages of Europe the mood changed dramatically and the cat was savagely persecuted by the Christian Church. For centuries its association with witchcraft and black magic gave it an air of mystery and caused it to suffer endless torment at the hands of the pious. It was the sentimentality of the Victorians in the nineteenth century that finally rescued it and restored it to the role of a loved household pet. Since then, in modern times, its popularity has risen steadily. Today there are about six million cats in Britain. Of these five million are well cared for as companion animals in the home. One million are strays, managing somehow to survive in country fields and city alleyways.

Everywhere the cat numbers are multiplying, but nowhere so sharply as in the United States. A survey there a few years ago showed that there were 35 million cats and 48 million dogs. But a further investigation in 1987 revealed that the totals had risen to 56.2 million cats and 51 million dogs. In other words, the increase in the number of dogs has merely been keeping pace with the growth in the human population, but the figure for cats has risen dramatically and for the first time they outnumber 'man's best friend'.

American sociologists have been intrigued by 'The Attack of the Cat People', as one Californian writer headlined the upsurge in addiction to felines, and have been searching for explanations. Some see it as a reflection of the number of owners who have to go out to work; or the increase in lonely people living by themselves; or the modern reluctance of many young couples to have children too soon. Others stress the rise in numbers of elderly people in the population, or the further shift towards apartment-dwelling.

In all these cases, cats have advantages over dogs. The

house-pets of working owners must be content with hours every day spent on their own, and cats do not mind being left alone in the home, whereas dogs crave non-stop company. Dogs need more space, so apartment-dwellers also find cats more suitable in the home. And dogs are always pleading for long walks, something that the elderly cannot always give them, much as they would like to.

These factors may explain most of the rise in feline popularity, but some observers feel that there is an additional, hidden influence that has to do with the changing social mood of recent years. They detect a growing respect for individualism and a decline in mindless loyalties to our major institutions. Respect for authority, they say, is more and more difficult to drum up with the usual slogans and ceremonies. It is still there, of course, but with it there is a cynical tinge – a wry smile that says 'we will follow you, but in our own way, and really we cannot take you too seriously'. The politicians, priests and pundits may still be our social leaders, but now, instead of looking up to them, we spend more time looking down at their clay feet. All of this sounds much more like a feline relationship than a canine one. Dogs are loyal slaves, ready for action to aid the pack. Cats are independent and individualistic, tolerating our leadership but not respecting it. They therefore reflect the new social mood of the human community and our co-existence with them suits that mood perfectly. We admire in them what we feel growing in ourselves.

Connected with this change is a significant shift in animal literature. Quirky cats have replaced heroic dogs in childrens' stories to a striking degree. And it is *Cats* not *Dogs* that are starring on stage at present in the West End of London and Broadway in New York.

It is a sobering thought that in the United States the figure for cat food sales, which has doubled in the last eight years to an astonishing $2 billion a year, now exceeds that for baby food. In weight, Americans are now buying 2.25 billion pounds of cat food each year – not to mention a million tons of cat litter. There is no doubt about it – we are clearly living in the Age of the Cat.

Index